Michael Friedman and Wolfgang Schäffner (eds.)
On Folding – Towards a New Field of Interdisciplinary Research

Michael Friedman and Wolfgang Schäffner (eds.)

On Folding

Towards a New Field of Interdisciplinary Research

[transcript]

This publication was made possible by the Image Knowledge Gestaltung.
An Interdisciplinary Laboratory *Cluster of Excellence at the Humboldt-Universität
zu Berlin with financial support from the German Research Foundation as a part
of the Excellence Initiative.*

Sponsored by

Bibliographic information published by the Deutsche Nationalbibliothek
The Deutsche Nationalbibliothek lists this publication in the Deutsche Nationalbiblio-
grafie; detailed bibliographic data are available in the Internet at *http://dnb.d-nb.de*

Cover layout: Kerstin Kühl
Cover illustration: Garry Killian
Layout and Typeset: Kerstin Kühl
Copyediting: Orit Davidovich (Introduction, Seppi, Ferrand et al., Friedman et al.,
Gholami et al.), Benjamin Carter (Krauthausen, Schramke), Martina Dervis (Blümle)
Printed in Germany
Print-ISBN 978-3-8376-3404-4
PDF-ISBN 978-3-8394-3404-8

Contents

Michael Friedman, Wolfgang Schäffner

On Folding: Introduction of a New Field of Interdisciplinary Research

I.

Folding "is at work everywhere,"[1] it bends and weaves, it manifests and creates, it nourishes and is nourished, it operates materially and materializes operationally. Nature, and, one might claim, organic matter, is modulated through and impregnated by folding: one might wonder whether there is a limit to folding within and outside the organic realm of matter, since it seems to involve all scales, from the largest – as with inorganic geological folding processes – to the smallest – as with the DNA molecule.

Folding can be taken as a materialized operation: some frogs fold leaves to secure their eggs,[2] while chimpanzees fold leaves to swallow them whole.[3] Some trees fold their leaves while it rains or after certain hours.[4] As a spatial operation, folding can occur in at least three different dimensions. One-dimensional linear folding of fibers, two-dimensional planar folding of strata, leaves or surfaces, and three-dimensional folding of bodies can all be considered as various, spatial and architectural layers of matter as well as geometrical processes of folding.[5] Back to the most basic of organic levels, the protein, the brain or the embryo, as what one might call the beginning of life or its basic units, materialize operationally through and by folding and folds. Proteins fold themselves

---

1   Deleuze/Guattari 1984, 1: "Ça fonctionne partout, tantôt sans arrêt, tantôt discontinu."
2   For example the waxy monkey tree frog. See Wells 2007, 47.
3   Huffman et al. 2013, 20.
4   See e.g. the tree Albizia saman. See also: Satter et al. 1974, 413.
5   O'Rourke and Demaine also approach folding from the perspective of different dimensions: "The objects we consider folding are 1D linkages, 2D paper, and the 2D surfaces of polyhedra in 3-space." In: Demaine/O'Rourke 2007, xi.

with amazing speed, which begs the question: how does protein 'know',[6] out of million possible folds, to choose the right one so quickly?[7] The embryo and the brain materialize themselves only through the process of being folded adaptively.[8] The fold, we claim, is a material operation or/and an operative material starting at the molecular and ending at the macro level.[9]

In human culture folding plays an essential role as well. It is found in techniques of textile manufacturing, such as weaving, knotting and braiding, in calculating[10] and more importantly in cultural techniques of writing. Each technique might be thought as dealing with a sequential series of signs. Note, for example, Alexandre-Théophile Vandermonde's 1771 system of notation for knots, arising directly from folding, weaving and the textile industry,[11] or consider the origins of computer-science as having its roots in weaving: Basile Bouchon's 1725 punch-card controlled loom further improved on by his assistant Jean-Baptiste Falcon as a precursor to the highly successful Jacquard loom of 1805.[12] In addition, the mathematician Carl Friedrich Gauss first attempted to find a formalization the braid: a collection of folded, interweaved within each other curves formulized through sequence of letters.[13] Examining the techniques of writing, papyrus was one of the ancient mediums for linear alphanumeric systems – a rolled surface. Other methods for storing information existed before and alongside paper and papyrus, for example, clay tablets,[14] knots in threads or strings (e.g. Quipu used for administrative and calculation purposes) and hieroglyphs carved into stone.[15] But looking closely at paper and papyrus, linear writing was incorporated within and enabled by the papyrus's two-dimensional rollable surface. Papyrus had to be unrolled for writing and reading, whereas storing and transferring was

---

6   Here, the following question is also called for: is protein an agent, having in its possession implicit knowledge, or should one expand the agent of the *Savoir-faire*?

7   This is known as the *Levinthal's paradox*, a thought experiment conceived in 1968. See Levinthal 1968.

8   Susan 2007, 72: "As new cells form, the embryo is forced to conform to available space. To adapt to the confined space, the embryo folds (curves) in both transverse and longitudinal planes. Folding in the transverse plane causes the embryo to become cylindrical in shape; longitudinal folding results in the head and tail folds. Structures within the embryo (e.g., heart, intestines) also undergo folding to conform to the space available to them." See also Marín-Padilla 2011, 11.

9   See Seppi's contribution in this anthology concerning the logic the fold beckons towards.

10   Cf. also Krämer/Bredekamp 2003, 17, where calculation, although stemming from particular cultural techniques, is described as a "forgetting machine."

11   See Epple 1999, 52, where Epple describes Vandermonde's invention as follows: "The mathematization of textile pattern dealt also with the rationalization through standardization" (translated by the authors). Epple links standardization in the textile industry to the standardization of notation of pattern.

12   See Schneider 2007, esp. chapter IV (system of notation for textiles) and chapter IX, regarding punch cards.

13   Epple 1998. Cf. also Epple/Krauthausen 2010, 131f. concerning the "Papiertechniken." See also Epple's treatment of the mathematization of the braid (Epple 1999, chapters 6.3 and 6.4). Another origin of the mathematical study of braids began with the investigation of the simultaneous movement of a finite number of points in the complex plane (ibid, 185–192).

14   In Mesopotamia, cf. Glassner 2003.

done while rolled. The rolling of papyrus produces wrinkles without creases, thus rolling and unrolling served a perfect reversible way of folding and unfolding text.

Compared with papyrus, paper is a highly foldable material and therefore able to incorporate folding processes such as creases. With the introduction of the codex, the spatial organization of text changed from that of a continuous roll to that of a sequence of pages, which are connected through the spatial orientation of the sheet. For this reason, Immanuel Kant took the sheet of written paper as a starting point for his epistemological analysis of space and geographic orientation.[16] The written line has an orientation in space, which changes as the page is turned, rotated, bent or folded. In Kant's time, a sheet of paper could have also meant a book page, printed paper that is incorporated in the complex spatial arrangement of printed sheets.

The alphanumeric code – its writing and storage – is thus deeply linked to the material operation of folding. Taking the fold as an epistemological unit, as Gottfried Wilhelm Leibniz did, shows that it is nothing but a basic one. The fold functions neither as a basic element (the indecomposable chemical element), nor as one of the composing elements of Morse code (the line and the dot). Indeed, it does give rise to other operations and elements, but one cannot point to the manner in which this unit has originated. This is where the epistemological discourse concerning folding begins. The geometric point, defined by Euclid as the basic element among all spatial elements, is for Leibniz no longer a minimal or static object, but an operation of folding.[17] Not only is the fold already there – manifest in the protein, the brain or the embryo – it is always in-between. There is always a movement from one fold to the next, refusing a reduction to the basic units of life:

> "[e]verything moves as if the re-folds [*replis*] of matter possessed no
> reason in themselves. It is because the Fold is always between two
> folds, and because the between-two-folds seems to move about

---

15 Another controversial narrative regarding the development of writing from calculation is given by Denise Schmandt-Besserat, regarding the move from three-dimensional clay tokens to their two-dimensional negative imprint on their containing envelopes. According to Schmandt-Besserat, this move occurred in order to preserve the function of the three-dimensional clay tokens, by sealing them in hollow clay spheres. Cf. Schmandt-Besserat 1992.

16 Kant 1991, 29: "In a written page, for instance, we have first to note the difference between front and back and to distinguish the top from the bottom of the writing; only then can we proceed to determine the position of the characters from right to left or conversely. Here the parts arranged upon the surface have always the same position relatively to one another, and the parts taken as a whole present always the same outlines howsoever we may turn the sheet. But in our representation of the sheet the distinction of regions is so important, and is so closely bound up with the impression which the visible object makes, that the very same writing becomes unrecognizable when seen in such a way that everything which formerly was from left to right is reversed and is viewed from right to left." See also Krauthausen's contribution regarding how folding in books subverts the notions of dimensionality and directionality.

17 This conception expressed for example in: Deleuze 1993.

everywhere: Is it between inorganic bodies and organisms, between organisms and animal souls, between animal souls and reasonable souls, between bodies and souls in general?"[18]

Beyond the two folds, the materialized operation and operative matter, let us briefly examine one historical cultural narrative of folding as reflected in origami. The history of origami comprises two origins: western and eastern.[19] The origin of paper folding in Japanese tradition is rooted in ceremonial packaging. It goes as far back as the 17th century or possibly earlier. The western tradition of origami can be traced back to 16th century German baptismal certificates folded into a *double-blintz*. Folding is also evident in Albrecht Dürer's 1525 manuscript *Underweysung der Messung mit dem Zirckel und Richtscheyt*, in which he suggested tracing two-dimensional drawings of unfolded polyhedra, then cutting them out to construct the finished polyhedral.[20] However, it was Friedrich Fröbel, the German inventor of kindergarten, who contributed most to the development of folding throughout Europe and beyond during the 19th century. Interestingly, in 1814, Fröbel worked as an assistant in the mineralogical museum of the University of Berlin. He assisted in the classification of the museum's crystal collection, while taking classes from Christian Samuel Weiss on crystallography and mineralogy.[21] Maintaining that the fixed laws of nature also guide the world of infants and adults, it is therefore obvious that Fröbel's ideas already contained a geometrical mathematical kernel, to be found in the spatialization and the mathematization of folding. Indeed, starting from the second half of the 19th century, one finds several school and kindergarten instructional materials referring to folding.[22] Although Fröbel did not consider paper folding a mathematical operation, it has certainly inspired such approach, culminating in Tandalam Sundara Row's work.[23] In it, folding is presented as having a purely mathematical basis: once one builds the basic folds, other folds can be derived from the already existing system. It is equally important to note that Fröbel links two-dimensional folding of paper directly to one-dimensional folding present in weaving and braiding and also to three-dimensional arrangements of building blocks. Fröbel thus shows how these different fundamental cultural practices converge in the material and spatial operation of folding. Fröbel's *Kindergarten Gifts* demonstrate it as a geometrical exercise.

---

18   Deleuze 1993, 13 (translation changed).
19   We follow the historical account presented in Hatori 2011 and in Lister 1997.
20   Dürer 1525. See also Heuer 2011. Note that during Dürer's time paper folding was not considered at all a mathematical object. However, Dürer's work directly influenced the mathematicians that followed him: Wolfgang Schmid and Augustin Hirschvogel (cf. Richter 1994, 58 – 66). This method had possibly influenced Henry Billingsley's 1570 translation of Euclid's *Elements*, adopting Dürer's diagrams and folding suggestions, and also 1613 Denis Henrion's description of the folding of platonic solids, in his *Memoires Mathematiques*.
21   Cf. Lilley 1967, 15.
22   For example: Miller/Macaulay/Stevens 1855; Martin 1893.
23   Row 1893.

The two traditions of folding – the eastern and the western – merged towards the end of the 19th century, due to the introduction of Fröbel's system into Japanese kindergartens concomitant with Japan's opening of its borders to the western world. During the 20th century further development took hold: a standard notational system was accepted, using the one established by Akira Yoshizawa supplemented by Randlett-Harbit. With standardization there comes the possibility for an algorithm: a specific set of instructions for carrying out a procedure, in this case, the final folded form. However, what is clear from this brief overview is that origami and folding were not always conceived as codified system, composed of basic folds, basic units; for a long time it was not even considered as such.[24]

The idea that folding should not necessarily be codified is also reflected in the growing understanding of the structure of proteins and DNA – one of 20th century's most important fields of study where folded structures emerge. Even before the discovery of the double helix, molecular biologists, Alfred Mirsky and Linus Pauling, described proteins in 1936 in terms of a folded structure.[25] Folding was tightly connected to form in matter. A fundamental association between folding in organic matter (such as in proteins) and alphanumeric code is evident already in Schrödinger's famous book *What is life?* based on a lecture he gave in 1943.

Schrödinger relates code to aperiodic crystals, and offers to think of chromosomes as coded. However, he adds that

> "[...] the term code-script is, of course, too narrow. The chromosome structures are at the same time instrumental in bringing about the development they foreshadow. They are law-code and executive power – or, to use another simile, they are architect's plan and builder's craft – in one."[26]

It is obvious that, if indeed so, folded genetic material is a very special code, compared with alphanumeric and digital codes, since it not only encodes different activities but also operates as a type of motor and designer at the same time. In this sense, the fold opens the possibility for a new form of analog coding; but not as one of several types of analog codes,[27] rather in a general sense, which reflects the relations between matter

---

24 It is interesting to note that the axiomatization of geometry based on folding took place only in 1989 independently in Justin 1989 and Huzita 1989.

25 Mirsky / Pauling 1936.

26 Schrödinger 1944, 22.

27 Analog codification, in contrast to digital codification (as a discrete representation or symbolization of information, e.g. the Morse code or the Braille system), is a continuous representation of codified information enabled by a continuous set of values (for example, voltage signals). It employs different mathematical tools compared to those of digital codes, namely for example, convolution and Fourier transform.

and code. Schrödinger's point of reference is not digital code but Morse code, when he reflects on the relationship between the number of signs and their possible combinations within the "miniature code [...] of this tiny speck of material,"[28] that is, when he examines the relationship between code and matter.[29]

It seems therefore that, although folding is one of the basic processes and operations in existence, it could be regarded as encoded. Nonetheless, can one say folding is or could be coded? What are the relations between folding, folded materiality and code? Is the fold a necessary constitutive condition or just an after-effect? It is here that the question of code, especially digital code, arises as what opposes folding. Digital code is normally understood as what codifies operations and processes into an alphanumeric series of signifiers, enabling us to view operations – such as folding, but also and obviously, text – as a linear series of transmissible, discrete operations,[30] which can be repeated over and over.[31] Starting from the end of the 19th century, code was no longer perceived as what stems from a codex, but rather as what codifies and externalizes thought or meaning[32] through the codification of difference itself.[33] Transmitted alphanumeric code points towards deciphering, reading and writing as what belongs to digital one-dimensional code, as we shall see later on.[34] Digital code is ultimately denuded of any sign of materiality; it represents pure form overcoming not only materiality but also the surrounding material environment and space. Folding, we claim, represents fundamentally nothing of this sort. Might it be said that digital code, comprising distinct separable elements, represses the constitutive aspect of folding?

---

28 Ibid, 61. See also: "What we wish to illustrate is simply that with the molecular picture of the gene it is no longer inconceivable that the miniature code should precisely correspond with a highly complicated and specified plan of development and should somehow contain the means to put it into operation" (ibid, 62).

29 See Kay 2000, 59 – 66. Examining Schrödinger's lecture, how it was conceived and historiographized. Kay claims in addition that Schrödinger did not think on information or on "one-dimensional Boolean message inscribed on a magnetic tape" (ibid, 66).

30 Cf. Kittler 2008.

31 Cf. Deleuze 2003, 114.

32 See Slater, 1888, v – xvi: the sentence "The Queen is the supreme power in the realm" is coded via sentences whose words have no relations to each other. One can detect this externalization in Shannon's Master thesis (Shannon 1938), where the simplification of electrical circuits is done via Boolean algebra, i.e., as what operates automatically. Concerning this conception of Boolean Algebra, cf. Schäffner 2007. It is crucial to note that Slater 1888 refers to code as what "ensures secrecy" (Slater's 1888 book title), hence there is a transition from the codex as known-to-all to code as barely known to two individuals: sender and receiver.

33 The codification of difference in language can be seen in Morse code in the visual difference between '–' and '•'. Compare also Deleuze 2003, Chapter 12. Deleuze indicates that the diagram, composed of "asignifying traits" being neither signifiers nor significants (ibid, 100), is replaced by digital code.

This book aims to reshape the too-well-known narrative that locates folding under the category of digital code and hence allows it to disappear.[35] We wish to not merely suggest a post-digital conception of coding, but to point towards its possible incarnation through and by folds and folding. This already requires different conceptions and possibilities of codification based on an understanding of a local notion of code.[36] We attempt to do so by proposing folding as a fundamental operation that not only preceded digital code, but has also shaped it; in this context digital code emerges as a reduced form of folding. Were it to emerge as such, a reconstitution of analog code would be feasible as well. The relationship between the continuous and the discrete would have to be reshaped.

II.

Before we embark on our program, let us (presumably) step back from the cultural sphere. Let us step back from the narrative, depicting folding in its infancy as a primitive cultural technique 'progressing' towards a codification, and consider the study of protein folding during the 1950s. Towards the end of 1956, in a symposium on information theory in biology, RNA was described by the Lithuanian microbiologist Martynas Yčas as follows: "[...] the RNA molecule can be regarded as a text, written in a four-symbol alphabet, which encodes another text, the protein, written with about twenty symbols."[37] One might therefore assume the folded structure of these molecules can be coded fairly easily, as Yčas indicated: "Since both DNA and RNA are texts written in a four symbol alphabet, it is natural to suppose that the coding problem is very simple."[38] As hopeful as his prediction was, Yčas followed with: "[r]ecent evidence indicates, however, that this is incorrect."[39]

---

34 Here we refer to Jakobson 1971, 276–278, where the problem of understanding language is turned (in the context of Markov chains, information theory and probability) into a problem of deciphering and analyzing code with its distinctive features. Compare also Cherry/Halle/Jakobson 1953 and also Halle/Jakobson 1956, 5.

35 For an analysis of the understanding of folded structures as what can and should be codified within biology see Kay 2000, chapter 2, where Kay describes the move from a discourse of organization and specificity to a discourse of information: "Specificity [...], grounded in three-dimensional molecular structures, [interchanged to] Information [...], abstracted as a one-dimensional tape, a transaction devoid of experimental measures and material linkages" (ibid, 41).

36 Concerning the locality of the code, we note that, before the Mesopotamian sexagesimal place-value system evolved into the chief method of number representation in the 19th century BC, different systems of numerals were used in various contexts (Damerow 1996, 161–162), which means that the use of code was local, i.e. context-dependent.

37 Yčas 1958, 70.

38 Ibid, 89.

39 Ibid.

Why was Yčas's conception of coded DNA doomed to fail? Indeed, one might claim the above symposium is the 'wrong' epilogue to Schrödinger's codification program, as it sought to reduce RNA to the form of digital code, instead of exploring the possibilities afforded by the aperiodicity of code. As Schrödinger saw it – suggesting the interlacing of weaving and code – compared to a repeating pattern, codification was envisioned a "masterpiece of embroidery."[40] Yčas's concluded that through reduction to codification the problem of RNA structure and function becomes "very simple". Reduction makes it easy to resolve, but then Mirsky and Pauling's folded structure of proteins does not have real implications for codification. In the midst of the rise of the information technology industry, this symposium had powerful consequences regarding the discourse of folding and codification.

It is important to note that, towards the end of his paper, Yčas offered the following solution to the above problem: "This suggests that a whole series of codes of this type may exist, all having similar general properties. At present the major difficulty is not to produce a coding principle that explains the known facts, but rather to make a choice between the many that are possible."[41] The solution is not to abandon digital code but rather to stick to it. This viewpoint was expressed in a 1965 lecture by the biologist François Jacob at the College de France, where he stated that a chemical message is written not with what is drawn, but rather with alphabet, or more precisely using a code such as Morse code.[42] This corresponds to Yčas' suggestion: either a multitude of codes or a search for a better code, eventually using "the electronic computer."[43]

Here lies the problem in the conception of digital code in the study of RNA/DNA. Greatly affected not only by the Wiener-Shanon discourse but also by the invention of the *Turing machine*, the perspective that emerged at the symposium was that computation is a linear operation.[44] Alan Turing's seminal paper, *On computable numbers*, is considered one of the breakthroughs of modern computer science.[45] The paper describes a model for every computing machine and solves in addition Hilbert's *Entscheidungsproblem*.[46]

---

40 Schrödinger 1944, 5.
41 Yčas 1958, 93.
42 Jacob 1965, 22: "Heredity is determined by a chemical message inscribed along the chromosomes. The surprise is that genetic specificity is written, not with ideograms as a Chinese, but with an alphabet as in French, or rather in Morse. The meaning of the message comes from the combination of words and signs in the arrangement of words into sentences. The gene becomes a phrase of a few thousands of signs, started and terminated by a punctuation" (translation of the authors). Compare also Liu 2010, esp. chapter 2.
43 Yčas 1958, 93.
44 "There is good reason to believe that machines will actually be built which can compute any number that can be computed, and which, even more generally, can arrive at the results of any thinking which can be described by explicitly-defined operations" (Quastler 1958, 7).
45 Turing 1936.

The greatest achievements of the Turing machine is arguably not in providing the proper model for a machine, but rather in supplying a definition of computability, that is, which numbers can be computed *de facto*.

Computation was done for centuries by hand, either literally, as with the abacus, or through writing, when more complicated computations demanded it. Computational apparatuses predated the Turing machine, nevertheless Turing's idea can be thought of as replacing the writing-computing hand with a mechanical process, independent of direct human intervention. However, in this exchange, the topological surface of computation was left out, much more than the human agent.[47] The two-dimensional object, the paper, as it is embedded in a three-dimensional space, was forgotten.[48]

The writing of the Turing machine is considered to be denuded of any trace of materiality. In its description, an emphasis was placed on the recording of transitions from one state to the other and certainly not on the physical materiality of the Turing machine: "A computable sequence γ is determined by a description of a machine which computes γ."[49] Indeed, computation was reduced to machine writing, based on an axiomatic system.[50] When Turing provided an account of the workings of a computer, he stated its operations could be reduced to elementary steps: "Let us imagine the operations performed by the computer to be split up into 'simple operations' which are so elementary that it is not easy to imagine them further divided."[51] In a striking parallel to Hilbert's famous remark on geometry as an axiomatic system,[52] the relation of machine operations to their actual realization was disregarded in favor of the relations between the elementary operations. In contrast to Hilbert's approach, who still made an effort to draw two-dimensional diagrams in his famous book *Grundlagen der Geometrie*,[53] Turing based his machines, and

---

46 Hilbert's *Entscheidungsproblem* asks whether there is a general process for determining if a given statement A is provable, i.e. whether there is a machine which, supplied with A, will eventually say whether A is provable, i.e. deduced from the axioms of the logical system (cf. Ackermann/Hilbert 1928, chapter 3). Turing answered in the negative (Turing 1936, 259).

47 Cf. Deleuze 2003, 104, when replacing diagram with code "[t]he hand [of the painter] is reduced to a finger [*doigt*, in French] that presses the" digits.

48 See also Friedman/Krausse's contribution to this anthology, regarding Buckminster Fuller's critique of Euclid's ignorance of the forgotten surface.

49 Turing 1936, 239.

50 Ibid, 230: "a number is computable if its decimal can be written down by a machine."

51 Ibid, 250.

52 It is important to remember that although Hilbert's views regarding geometry in particular and mathematics in general were not mathematics as an empty formal game (a view which is exemplified in his book *Anschauliche Geometrie*, Hilbert/Cohn-Vossen 1932), he stated, on his return from Halle after hearing Hermann Wiener's lecture, his famous expression, "One should always be able to say, instead of 'points, lines, and planes', 'tables, chairs, and beer mugs'" (Blumenthal 1935, 402–3).

53 Kittler advocates this approach: "the entire foundational crisis of mathematics boiled down to the question whether numbers 'exist in the human mind', [...], or 'on *paper*', as [...] Hilbert was able to convince his contemporaries and posterity" (Kittler 2006, 47, our italics).

hence, the computability of code, on one-dimensional linear representation: "I assume then that the computation is carried out on one-dimensional paper, i.e. on a tape divided into squares."[54] Turing even allowed for an infinite strip of paper. One may understand this continuous infinite strip as analog code that enables the digital code written on it.[55] Nevertheless, this paper machine still does away with any sign of materiality associated with the paper strip and consequently virtualizes it.[56] This influential understanding of a computing machine as a paper machine paved the way to the understanding of codification as a linear process, dealing with the transmission of information, in a way which can be emptied not only out of materiality,[57] but also out of any form of spatiality.[58] Linearity henceforth is thought of in terms of virtual linearity; it does not assume any dimensional character, as it is not embedded anywhere – spatiality is removed.

This symbolic linearity is the basis for the linearity of code; it establishes the allegedly successful encounter between biology and information theory. At this point, the French mathematician René Thom vehemently attacked the conception that biology deals with the transmission of information and should be understood in terms of a linear code:

> "I discussed [in section 7.2] [...] the abuse of the word 'information' in biology; it is to be feared that biologists have attributed an unjustified importance to the mathematical theory of information. This theory treats only an essentially technical problem: how to transmit a given message in an optimal manner from a source to a receiver by a channel of given characteristics. The biological problem is much more difficult: how to understand the message coded in a chain of four letters of DNA (the usual description of which, it seems to me, needs some improvement). [...] to reduce the information to its scalar measure (evaluated in bits) is to reduce the form to [of] its topological complexity [...], and to throw away almost all of its significance."[59]

---

54 Turing 1936, 249.
55 Cf. Derrida 2001, 20. Remarking on his artisanal writing and paper machines, Derrida says the following: "As if that liturgy for a single hand was required, as if that figure of the human body gathered up, bent over, applying, and stretching itself toward an inked point were as necessary to the ritual of a thinking engraving as the white surface of the paper subjectile on the table as support. But I never concealed from myself the fact that, as in any ceremonial, there had to be repetition going on, and already a sort of mechanization."
56 See also Dotzler 1996, 7.
57 To borrow Derrida's expression, there is a "de-paperization" of paper, i.e., of the support (Derrida 2001, 55).
58 We follow here Sybille Krämer, who wrote: "In this respect the result of the internal states of the machine, which are held by tablets, are not states of a concrete operating apparatus, but of sign configurations: this machine does not occupy any particular place in time and space, but only on paper" (Krämer 1998, 171, our translation). Departing from Krämer, we claim one might add 'virtual' paper, that is, with Turing, paper lost its concrete reference to matter.

Michael Friedman, Wolfgang Schäffner

What are the grounds for Thom's critique? Thom did not discard of the term 'code' as such, but rather called for its re-evaluation. Thom criticized Turing's conception of code. He spoke against the view that code is a priori linearized and alphanumeric, that one may read it as if it was linear text, as did Yčas, while ignoring the fact that objects in the world have a "topological complexity."[60] In the above citation, taken from section 8.4 of *Structural Stability and Morphogenesis*, Thom refers to section 7.2, there he expands on his notion of topological complexity. Thom remarks: "the [topological] complexity of F [a dynamical system] is rarely defined in a way intrinsic to F itself." Not only does the dynamical system lack intrinsic codification, but one cannot also impose an extrinsic codification without considering the family in which the system is embedded.[61] Indeed, Thom states:

> "If, as Paul Valery said, 'Il n'y a pas de géométrie sans langage,' it is no less true that there is no intelligible language without geometry, an underlying dynamic whose structurally stable states are formalized by the language."[62]

Considering the fold as what represents topological complexity, Thom does not propose to codify folding by imposing on it a single language or one-dimensional linear codification.[63] Instead, he proposes to fold code. What is necessary for digital code to emerge is geometry, or in Thom's words a "topological complexity", an underlying dynamic, which is not reducible to its "scalar measures." At best, linear code should be viewed as a reduced form

---

59 Thom 1975, 157. Let us note that in the last sentence Thom uses the preposition 'to' ('à' in French), whereas we suggest a more proper proposition would be 'of'. The meaning implied by 'to' is that a reduction to "bits" enables an intrinsic reduction of topological complexity. However, as footnote 62 suggests, Thom (in Section 7.2 in Thom 1975) advocates another approach (which is implied also from using the proposition 'of').

60 A similar criticism was directed at Lily Kay's book *Who Wrote the Book of Life? A History of the Genetic Code* (Kay 2000). Against the erroneous depiction of DNA as "self-replicating," as "making" proteins and as "determining organisms," Lewontin offers that "organisms are not determined by their DNA but by an interaction of genes and the environment, modified by random cellular events" (Lewontin 2001,1264).

61 Thom 1975, 127: "The scalar measures [...] should be geometrically interpreted as the topological complexity of a form. Unfortunately it is difficult, with the present state of topology, to give a precise definition of the complexity of a form. One difficulty arises in the following way: for a dynamical system parameterized by a form F, the complexity of F is rarely defined in a way intrinsic to F itself, for it is usually necessary to embed F in a continuous family G, and then define only the topological complexity of F relative to G." Hence, it is not that a geometrical form should be defined and reduced via its scalar measures, but rather the other way around.

62 Thom 1975, 20.

63 Thom stands against Husserl's conception, presented in Husserl 1989, that the Pythagorean theorem is identical in every language: "The Pythagorean theorem, [indeed] all of geometry, exists only once, no matter how often or even in what language it may be expressed. It is identically the same in the 'original language' of Euclid and in all 'translations'; and within each language it is again the same, no matter how many times it has been sensibly uttered, from the original expression and writing down to the innumerable oral utterances or written and other documentations." (Ibid, 160).

of folding, and not the other way around. A merger of materiality and operationality – for instance, in the form of 'virtual' paper – already informs Turing's codification of computation.

III.

Digital code is realized by means of writing, reading, inscribing, copying: simple operations – a term that appears in Turing's and Yčas's work – arranged in a virtual line (as is also the case with writing, reading and so on). As these are the terms used to describe code as a linear procedure, what should then be the terms used to discuss folding?

First and foremost, it is instructive to recall the following simple fact: folding is a spatial operation. The line that appears as a result of folding a piece of paper is a spatial *effect*, an after-effect of folding paper into a three-dimensional shape. Second, we suggest that a folded code should involve additional operations which incorporate three-dimensional folding processes that go beyond the mere sequential chain of equally-connected symbols, to include three-dimensional operations, such as bending, stretching, twisting and translating.[64] Folding, as a constitutive structure, reestablishes the materiality of code and codification in three respects. First, it suggests that codification should be communicated through physical and material effects. Second, it transcends the dichotomy of the codifying symbols, namely, that these symbols can *either* represent objects or execute actions. A folded code can do both.[65] The above series of three-dimensional operations is certainly not comprehensive. In contrast to the former list of simple operations, which presupposes a priori one-to-one correspondence between code and message, it hints at the possibility of a varying adaptive structure, which could possess a differential growth unique to the structure itself. Third, the materiality of bending, stretching or twisting does not refer to ideal operations, but rather to mechanics, to the manner in which material itself changes.[66]

---

64 'Translation' here is used in the sense of moving from one place to another. Note that Hermann Wiener, who wrote a short manuscript on folded models of the platonic solids in 1893, suggested also to explore the relations between these operations and the operation of half-turn, which is based on folding (*Umwendung*). See Friedman 2016.

65 One may propose that an evident example for a folded code is Gödel's theorem: Gödel proposes to codify the arithmetical system into itself (attaching to every mathematical symbol an integer). This codification causes an operation to be executed in the now-represented structure itself: it reveals the existence of the incompleteness theorem, proving that there is a sentence, which is neither provable nor impossible to prove in this system.

66 See Guiducci/Dunlop/Fratzl's contribution in this volume.

Michael Friedman, Wolfgang Schäffner

In order to exemplify this new series of operations, let us briefly review their emergence in philosophy, mathematics and biomaterials. We should emphasize we do not endeavor to produce terms that apply analogously to each. Instead, we wish to demonstrate how folding ties together materiality, logic and the symbolic.

Etymologically the English, German and French verbs *to fold, falten*, and *plier* are derived through Latin from the Greek *plékein* (to plait, to weave). In Hebrew the nouns derived from the verbs 'to fold' and 'to multiply' are homophones. This suggests Thom's viewpoint regarding the interweaving of language and geometry. Instead of pointing out how deeply language and thought are folded into one another, through etymological derivatives such as 'implication', 'explication' and 'application', it is essential here to turn to Deleuze-Leibniz: (un)folding is how thought interlaces itself and interlaced within itself.[67] Codification, on the other hand, does not aim to describe the interlacing of thought, material and operation. To borrow Alexander Galloway's expression in relation to his codification of Guy Debord's 1965 *Le Jeu de la Guerre*,[68] code has the "aesthetic of the superego: it mandates optimal material behavior through the full execution of rules."[69] Besides dematerialization, digital code demands totalization, as every word comes under the same symbolic system.[70] It produces an externalization of thought. As Galloway observes, the codification of Debord's game corrects and reveals players' mistakes[71] and hence irons out, one might say, the folds in thinking. As codification is but one instance of folding, folding that would cause thought to unfold itself, to move, borrowing Heidegger's terminology, between the *Einfalt* to the *Zwiefalt*.[72]

Unfolding in mathematics appears, for example, in Thom's work as a technical term (albeit not only as such). When a function f is singular, its unfolding consists of finding a family of functions having f as a member, where almost all other members of the family are not singular at all or less singular than f.[73] This might be considered a conceptual step forward: instead of studying a single function, one examines how the enveloping family of functions unfolds the singularity. In Deleuze's terms, this would mean concentrating on the space of problems, unfolded in front of us, instead of concentrating on axiomatics.[74] It is the problem that unfolds. Its solution solves nothing, and especially does not try to establish a system

---

67 Cf. Deleuze 1993, e.g. 31, 49
68 Becker-Ho/Debord 2006.
69 Galloway 2009, 148.
70 This is indeed the goal in Slater 1888: an allegedly complete dictionary of codification of all English.
71 Galloway 2009, 145.
72 Heidegger 2007.
73 See e.g. Arnold/Varchenko/Gusein-Zade 1985, 285–287. See also Ferrand/Peysson's contribution in this anthology.
74 To cite one example from Deleuze's work: Deleuze/Guattari 1987, 362ff.

of basic, irreducible units, but rather to move from "one fold to another."[75] Indeed, while unfolding a singular function, one might obtain a *bifurcation set*, that is, a whole sub-family of degenerate singular functions, of whom *f* is but one member. The interplay between unfolding a singularity and the emergence of new singularities, which in turn call for additional unfolding, gives rise to a mathematical process, along the lines of Thom's description of the genesis and morphogenesis of life, as what "permit[s] it to create successive transitional regimes."[76] Unfolding and folding consequently combine into a process without origin or end.

Differential growth in biomaterials provides the last but certainly not least example. Growth of leaves, as in kale (a variety of *Brassica oleracea*), demonstrates what happens when a leaf grows faster at the rim compared with the center. Since there is only limited space for the leaf to grow, the edge of the leaf begins to bend and fold, eventually developing a fractal form.[77] Biomaterials research aims to explain growth mechanisms in sheets where folding occurs. The purpose is not to codify every aspect of growth, but rather to deal with a process, which although codified, is always in transition. As a diachronic process of differential growth, folding changes the synchronic relations between different elements in the leaf, causing yet more folding to take place.

IV.

In this volume we present a cross-section of state-of-the-art research into the concept of folding. This collection of papers ranges from physics to architecture, from biomaterials to philosophy and art, and from literature to mathematics. Each discipline exhibits folding as a constitutive operation, which does not allow itself be reduced to a mere collection of linear codes. This can be seen in the first four papers, which deal with, what one might call, the non-material emergence of folds of thought that are nevertheless material.

Karin Krauthausen's contribution (*Literary Studies, History of Science*) suggests the manner in which the physical action of folding (of book pages, scrolls or paper) points to a conceptual crossover of the notions of dimension, direction and continuity. Following in the footsteps of Jacques Derrida and Gérard Genette, as well as analyzing the etymological, historical and cultural roots of paper folding in the context of literature studies, this paper argues that the fold conveys what linear writing had previously dissolved, namely, another praxis of nonlinear multidimensional reading.

---

75 Friedman/Seppi 2015.

76 Thom 1975, 289.

77 Barbier De Reuille/Prusinkiewicz 2010, 2121. See also Guiducci/Dunlop/Fratzl's contribution in this anthology.

Angelika Seppi (*Philosophy*) follows in her paper the logic of folds of thought and thinking, which crosses, ignores and shakes the chains of classical logic. Unfolding Deleuze's thoughts, Seppi cross-references a series of thinkers: Didi-Huberman, Derrida, Leibniz and Lucretius, each of whom beckoning at the other. A non-linear web, a root system, is revealed before our eyes, an unfolded thinking, which unfolds the folds, traversing known und unknown dichotomies: matter and soul, body and cloth, 'and' and 'or'.

Claudia Blümle (*Art History*) examines the folds in the paintings of El Greco. Following hidden mathematical layers that El Greco points to, Blümle discovers that the fold does not try to constitute any space, imaginary or real, in the plane of the picture itself, but is rather occupied with subverting any linear thought. Through the folded structure of drapes and gazes, a Deleuzian vector field emerges in El Greco's painting. Blümle's careful research conveys affinities with Seppi's contribution, a research that unfolds the subversion that the fold leads to and is.

The contribution by Dominique Peysson (*Physics, Art*) and Emmanuel Ferrand (*Mathematics*) is an experiment both in materiality and in the mathematical theory of folding, showing that the fold is neither a mere metaphor nor an event in pure materiality. Performing an experiment with folded sheets of paper, they demonstrate that the fold does not act only as a metaphor but has always and already material consequences whether in literature or mathematics. Revolving around the enigmatic understanding of folds by the mathematician René Thom, their contribution hints towards Krauthausen's, regarding a folded narrative that depends on its materiality and objects linearity, and towards Seppi's regarding the Deleuzian conception of the fold.

Folded materiality points in several directions, two of which are explored in the following four papers. The first is architecture, presumably dealing with static structures. The second is physics and biomaterials research, dealing with the stable structures of matter. However, thinking through folding demonstrates these disciplines deal with nothing similar to stability or stasis.

Sandra Schramke's contribution (*Architecture*) concerns the architect Peter Eisenman, who during the 1990s invented a new architectural language with the shape of the fold. Eisenman's work is based on Gilles Deleuze's writings and René Thom's chaos theory. Eisenman understands the fold as a form that builds a bridge between an inner and an outer world, between geometry and a mental state. The fold is based on the geometry of the grid, on perception as well as mathematical theories. Furthermore, the form of the fold implements models from the natural sciences, but without compromising the singular space of perception, which honors the Deleuzian event, inevitably connected with the fold. Eisenman's reflections regarding the fold deal with American neo-pragmatism, which led postmodernism from a dead end, in order to fulfill a new autonomy and singularity of aspirations of architecture.

Joachim Krausse (*Architecture*) and Michael Friedman (*Mathematics*) examine the multifaceted figure of Buckminster Fuller. Against the classical idea of stable structures of geometry, during the early decades of the 20th century, in the form of axiomatization, group theory and model theory, Fuller posits another view of material geometry, which demands its changing materiality in return. According to Fuller, geometry is not about points, lines and planes. Simply put, it is not about using discrete elements, but rather – in line with Gottfried Semper, Aaron Klug, opening seedpods and the Jitterbug dance – about continuous transformations, unfolding structures and movement: where all of these lead to a provocation of thinking, to an unfolding of it. Thought as both architectural and material provocation of movement sharpens the ideas raised in Ferrand/Peysson's contribution, concerning the fold as what arises from movement and is apparent only through it.

The contribution by Lorenzo Guiducci, John W. C. Dunlop and Peter Fratzl (*Bio-Materials Research*) describes the physical and biological transformations and processes that take place in leaves, wood, thin tissues and seedpods among others. They allude to Fuller's work, presented in Friedman/Krausse's paper, as processes that cause a never ending change in form: bending, folding, stretching and wrinkling. The folding of materials is not only an adaption of matter to a changing environment or a response to their own external codification, but is also a manifestation of thin tissues, as if the fold is already internally codified. The materials in question comprise multiple layers that interact with one another, subject to intrinsic nonlinear strains and compressions.

The final paper by Mohammad Fardin Gholami, Nikolai Severin and Jürgen P. Rabe (*Physics*) deals with folding of graphene and other carbon-based thin films. Indeed, as is already made clear in Guiducci/Dunlop/Fratzl, folding allows for the transformation of a two-dimensional material into more complex three-dimensional configurations. This paper tackles the question: how is self-folding encoded? How does one encode a continuous change?

V.

Let us attempt to summarize. Could the material operations of folding be considered a code albeit different from alphanumeric or digital codes? This is comparable to Schrödinger's thought concerning the relationship between chromosomes and code, a moment when code was introduced into biology. In this instance Schrödinger writes: "the term code-script is, of course, too narrow."[78] Folding thus helps us to rethink the classical dichotomy between continuous magnitudes and discrete elements, as already discussed in Aristotle's *Physics*: "whereas points can touch, discrete unities exist only as a series (ἐφεξῆς), since there is no in-between (μεταξὺ) of the symbolic elements."[79] Folding forces us to combine the continuous line and the discrete symbols into one and the same operation, while avoiding the conception of the digital as what withdraws from and denudes the underlying analog code. In the same way, one needs to develop a different concept of analog code, not as a transmission that varies over a continuous range with respect to sound, light or radio waves, among other mediums, but rather as a concept that emerges from the following three general characteristics of folding. These characteristics refer both to alphanumeric and digital codes as well as to the analog code, where both aspects, the digital and the analog, stem from folding as a materialized operation.

First, we argue that the totality of code – a totality present either by aspiring to discretely codify every word in the language or by conceiving any element as a part of a continuous analog code – is a conception that folding subverts.[80] Folding by its nature is local, a local adaptability. There is no infinite folding as there is no complete codification of material folded onto the material itself.[81] Folding is local, as it is adaptive to specific local conditions. An external total code, which codifies every material in the same way, ignores local conditions and adaptions that the code has to undergo. Even in the same material itself, the same structure, the same tissue, there are areas of different folded densities, whose code adapts differently.[82] In this sense, folding has an important relation to the whole. Surface tension defines the local fold and its being operational whereas changing environmental conditions generate continuous adaptions of the folding code.

Against this move towards a totality, evident both in digital as well as analog codification, we offer to formulate a conception of "minor sciences,"[83] which do not aim towards a

---

78   Schrödinger 1944, 22.

79   Aristotle 1961, Book V, Chap. 3, 227a 29 – 31, 96 (translation changed).

80   See Blümle's contribution to this volume, where one can come to understand the fall of folds as subverting the codification of the geometrical closed space.

81   Such an imaginary scenario is presented in Borges's *Del rigor en la ciencia*.

82   Cf. section 2.2.2 in Guiducci/Dunlop/Fratzl's contribution in this anthology. During a tree's growth, cells adapt to changing conditions (weight, temperature, moisture, etc.).

83   Deleuze/Guattari 1987, 361.

full uncovering of the underlying principles, but rather gestures towards the unfolding of an aggregate of problems,[84] an aggregate of local folded spaces spun together, each of which enabling a unique different codification. The codification of a growing folding leaf or an unfolded narrative[85] via differential geometry or via literary studies may intimate mutually adaptive folding codes, where their folded structures beckon to each other developing different areas of research.

Second, we claim that folding underlies a conception of materialized symbolic operations. Let us demonstrate that with two examples, already presented in brief: 1) looking at the holes of punch cards, one may observe that, though stemming from weaving, the holes exist now as a collection of discrete objects, not only detached from any act of weaving – from the action that suggested them – but also from any reference to reading. 2) Examining a folded line, created from folding a finite piece of paper, one may now neglect the paper and regard the line as the basic element – together with the infinite planes it may span – assuming an axiomatic approach, extended to infinity.

Against the background of the above two examples, let us take our cue from the architect Gottfried Semper, who claimed that the true manifestations of walls, that is, of space enclosures and separating planes, are woven materials: braids, mats, hangings and tapestries.[86] In light of this, we aim to bring back materiality into discrete elements – the point, the line, the plane – that is to say, we wish to relate back again the act of weaving and the point, or, more explicitly, the continuous and the discrete. Weaving itself is a synthesis of discrete elements – from the holes in the punch card to the creased points of the thread while being weaved – and continuous movement – the movement of the whole loom. We aim to bring back folded materiality as what opens up possibilities for an interplay of the continuous and the discrete: a hole is made possible by the continuous movement of thread, analog code is made possible through the digital discrete encoding of the knot, the thread or the molecule, through the discrete perforation of a punch card.

Third, we would like to point out that folding underlies not an externalization of thought but rather an internalization of a three-dimensional structure, a "topological complexity" that should not be lost or forgotten via any sort of codification. *Isomers* – molecules having the same chemical formula but different chemical properties – are the simplest example in this respect. Ever since Antoine Lavoisier (1743–1794), it was common to identify a material with its chemical composition. However, in 1828, Friedrich Wöhler discovered this is not enough, when he produced urea, which has the same chemical composition as ammonium cyanate (CH_4N_2O) albeit different properties. Jacob Berzelius coined the term 'isomer' several

---

84   Ibid, 362.
85   Cf. Krauthausen's contribution to this volume.
86   See e.g. Semper 1983, 21.

years later. This has led slowly to the understanding that "molecules had to be understood from the perspective of stereochemistry, that is, how atoms in a molecule are arranged in space relative to one another."[87] In 1848, Louis Pasteur discovered molecular chirality by investigating tartaric acid, thereby further emphasizing the importance of topological structure.[88] What both discoveries – Wöhler's and Pasteur's – indicate is the performative character – both executing and realizing – of a folded structure in comparison with a coded one. Hence, a discrete code – as in Lavoisier's codification of chemical elements in the shape of an alphanumeric code – is problematic. Although Lavoisier developed a material symbolic code, it nevertheless could not represent materiality adequately. The linear or diagrammatic symbolic representation had to be amplified by three-dimensional elements. Wöhler and Pasteur took on this crucial step, and this insight was later also prominent in 1874 van 't Hoff's discoveries.[89] In this respect, Lavoisier's codification may be considered an extreme reduction of folded code, achieved by ignoring its materiality and spatiality.

The change imposed on Lavoisier's codification, necessitated by new discoveries, as well as other aspects of folded code/coded fold, point towards a conception of an adaptive code. More than a local, symbolic and changing digital code, where a codifying letter has already several possible ways of execution, we seek an adaptive coded material, responsive to the feedback loop (from changing environmental conditions back to the material and vice versa), which is always operative, always changing the inner codification of the material and causes folds to emerge.

On this basis we suggest digital code be regarded an instance of a broader folded inter-lacing between analog and digital code. Thus, the conception of digital code could be revised. As with the Turing machine, the sequence of discrete symbolic states is only made possible via the continuous, local, material band, unrolled as an old papyrus, rendering the symbols readable, executable and realizable. Folding, as a material basis for symbolic operations, thus regains its importance as an amalgamation of analog operations and discrete symbols. It alters in addition the classical idea of digital code as a dematerializing process of a purely symbolic machine. When the folded band is viewed in terms of continuous algorithms and sequential operations that control their own execution, the old notion of analog code is rehabilitated.

---

87   Soledad 2008, 1201.
88   Pasteur 1848. Cf. Also Caillois 1973, 55, where the discovery is described in terms of the topological complexity ("La composition chimique des cristaux est identique, mais non leur topologie").
89   van 't Hoff 1877.

This points towards future challenges to come. If folding is "at work" everywhere, and folding can be regarded as an operation, coded in material and materialized in code, then a new analysis of folding is inaugurated. Folding of fibers and threads produces series of point-like folds, whereas the folding of plane surfaces results in lines, and folding of bodies in planes. Thus, the always-becoming structures of folding combine zero-, one-, two- and three-dimensional elements, as parts of an analog code that has to be developed in a broader sense. The idea of folding, as both opening up and necessitating new conceptions of an adaptive code, requires a program for developing new horizons for analog codes. Practical consequences of future perspectives on folding would follow the following three approaches: 1) an analytical approach applied to the development of the notion of folding as an analog code in mathematics, computer science, philosophy and media theory, 2) a historical and genealogical approach to analog code,[90] 3) an experimental approach within material sciences, biology and physics. Taking into account the dualism of matter and code would open up the analysis, history and experimentation of folding as a new field of interdisciplinary research.

---

90   Schäffner 2016.

Bibliography

Ackermann, Wilhelm / Hilbert, David (1928): *Grundzüge der theoretischen Logik*. Berin: Springer Verlag.

Applewhite, Philip B. / Geballe, Gordon T. / Galston, Arthur W. / Satter, Ruth L. (1974): *Potassium Flux and Leaf Movement in Samanea saman : I. Rhythmic Movement*. In: The Journal of General Physiology, vol. 64, no. 4, pp. 413–430.

Aristotle (1961): *Physics*. Trans. by Hope, Richard. Lincoln: University of Nebraska Press.

Arnold, Vladimir I. / Varchenko, Alexander / Gusein-Zade, Sabir Medgidovich (1985): *Singularities of Differentiable Maps: Volume I: The Classification of Critical Points Caustics and Wave Fronts*. Springer: Science & Business Media.

Becker-Ho, Alice / Debord, Guy (2006): *Le Jeu de la Guerre*. Paris: Gallimard.

Blackburn, Susan (2007): *Maternal, Fetal, & Neonatal Physiology*. St. Louis: Saunders.

Blumenthal, Otto (1935): **Lebensgeschichte**. In: Hilbert, David: Gesammelte Abhandlungen. Bd. 3: Analysis, Grundlagen der Mathematik, Physik, Verschiedenes, nebst einer Lebensgeschichte. Berlin: Springer, pp. 388–429.

Caillois, Roger (1973): *La Dissymétrie*. Paris: Éditions Gallimard.

Cherry, E. Colin / Halle, Morris / Jakobson, Roman (1953): *Toward the Logical Description of Languages in Their Phonemic Aspect*. In: Language, vol. 29, no. 1, pp. 34–46.

Demaine, Erik D. / O'Rourke, Joseph (2007): **Geometric Folding Algorithms: Linkages, origami, Polyhedra**. Cambridge: Cambridge University Press.

Damerow, Peter (1996): *Abstraction and Representation: Essays on the Cultural Revolution of Thinking* (Boston Studies in the Philosophy of Science, 175). Dordrecht: Kluwer.

Deleuze, Gilles (1993): *The Fold. Leibniz and the Baroque*. Trans. by Conley, Tom. London: The Athlone Press.

Deleuze, Gille (2003): *Francis Bacon: The logic of sensation*. Trans. by Smith, Daniel W.. London: Continuum.

Deleuze, Gilles / Guattari, Felix (1984): *Anti-Oedipus: Capitalism and Schizophrenia*. Trans. by Hurley, Robert / Seem, Mark / Lane, Helen R.. London. New York: Continuum.

Deleuze, Gille / Guattari, Felix (1987): *A thousand Plateaus: Capitalism and Schizophrenia*. Trans. by Massumi, Brian. Minneapolis / London: University of Minnesota Press.

Derrida, Jacques (2001): *Paper Machine*. Trans. by Bowlby, Rachel. Standford: Standford University Press.

Dotzler, Bernhard (1996): *Papiermaschinen*. Berlin: Walter de Gruyter.

Dürer, Albrecht (1525): *Unterweysung der Messung*. Nuremberg: Hieronymus Andrae.

Epple, Moritz (1998): *Orbits of Asteroids, a Braid, and the First Ink Invariant*. In: Mathematical Intelligencer, vol. 20, no. 1, pp. 45 – 52.

Epple, Moritz (1999): *Die Entstehung der Knotentheorie: Kontexte und Konstruktionen einer modernen mathematischen Theorie*. Braunschweig / Wiesbaden: Springer Vieweg.

Epple, Moritz / Krauthausen, Karin (2010): *Zur Notation topologischer Objekte: Interview mit Moritz Epple*. In: Krauthausen, Karin / Nasim, Omar W. (eds.): Notieren, Skizzieren: Schreiben und Zeichnen als Verfahren des Entwurfs. Zürich: Diaphanes, pp. 119 – 138.

Friedman, Michael (2016): *Two beginnings of geometry and folding: Hermann Wiener and Sundara Row*. In: BSHM Bulletin: Journal of the British Society for the History of Mathematics (in print). DOI: 10.1080 / 17498430.2015.1045700.

Friedman, Michael / Seppi, Angelika (2015): *Ein Dialog über die Falte: zwischen analogem und digitalem Code*. Lecture given as part of the yearly convention of the Interdisciplinary Laboratory Image Knowledge Gestaltung. Humboldt-Universität zu Berlin, 21 Novemeber.

Galloway, Alexander (2009): *Debord's Nostalgic Algorithm*. In: Culture Machine, vol. 10, pp. 131 – 156.

Glassner, Jean-Jacques (2003): *The Invention of Cuneiform Writing in Sumer*. Baltimore: Johns Hopkins University Press.

Hatori, Koshiro (2011): *History of origami in the East and the West before Interfusion*. In: Wang-Iverson, Patsy / Lang, Robert J. / Yim, Mark (eds.): Origami[5] Fifth International Meeting of origami science, Mathematics, and Education. Boca Raton, Florida: A K Peters / CRC Press, pp. 3 – 11.

Heidegger, Martin (2007): *Die Sprache*. In: id.: Unterwegs zur Sprache. 14th ed. Stuttgart: Klett-Cotta, pp. 9 – 33.

Heuer, Christoher P. (2011): *Dürer's folds*. In: RES: Anthropology and Aesthetics, no. 59 / 60, pp. 249 – 265.

Hilbert, David / Cohn-Vossen, Stephan (1932): *Anschauliche Geometrie*. Berlin: Springer.

Huffman, Michael A. / Nakagawa, Naofumi / Go, Yasuhiro / Imai, Hiroo / Tomonaga, Masaki (2013): *Monkeys, Apes, and Humans: Primatology in Japan*. Tokyo: Springer.

Husserl, Edmund (1989): *The Origin of Geometry*. Trans. by Carr, David. In: Derrida, Jacques: Edmund Husserl's Origin of Geometry: An Introduction. Lincoln / London: University of Nebraska Press, pp. 157 – 180.

Huzita, Humiaki (1989): *Axiomatic development of origami geometry*. In: Huzita, Humiaki (ed.): Proceedings of the First International Meeting of Origami, Science and Technology. Ferrara: Comune di Ferrara and Centro origami Diffusion, pp. 143 – 158.

Jacob, François (1965): *Génétique cellulaire, Leçon inaugurale prononcée le vendredi 7 mai 1965*. Paris: Collège de France.

Jacques, Justin (1989): *Résolution par le pliage de 'équation du troisième degré' et applications géométriques*. In: Huzita, Humiaki (ed.): Proceedings of the First International Meeting of Origami, Science and Technology. Ferrara: Comune di Ferrara and Centro origami Diffusion, pp. 251–262.

Jakobson, Roman (1971): *Zeichen und System der Sprache*. In: id.: Selected writings, vol. 2. Paris / The Hague: Mouton, pp. 272–279.

Jakobson, Roman / Halle, Morris (1956): *Fundamentals of language*. Gravenhage: Mouton & Co.

Kant, Immanuel (1991): *On the first ground of the distinction of regions in space*. In: Van Cleve, James / Frederick, Robert E. (eds.): The Philosophy of Right and Left. Dordrech: Kluwer, pp. 27–33.

Kay, Lily E. (2000): *Who Wrote the Book of Life? A History of the Genetic Code*. Stanford: Stanford University Press.

Kittler, Friedrich (2006): *Thinking Colors and / or Machines*. In: Theory, Culture & Society, vol. 23, no. 7–8, pp. 39–50.

Kittler, Friedrich (2008): *Code (or, How You Can Write Something Differently)*. In: Fuller, Matthew (ed.): Software Studies. A Lexicon. Cambridge: MIT Press, pp. 40–47.

Krämer, Sybille (1988): *Symbolische Maschinen. Die Idee der Formalisierung in geschichtlichem Abriß*. Darmstadt: Wissenschaftliche Buchgesellschaft.

Krämer, Sybille / Bredekamp, Horst (2003): *Kultur, Technik, Kulturtechnik: Wider die Diskursivierung der Kultur*. In: Krämer, Sybille / Bredekamp, Horst (eds.): Bild – Schrift – Zahl. München: Fink Verlag, pp. 11–22.

Levinthal, Cyrus (1968): *Are there pathways for protein folding?*. In: Journal de Chimie Physique et de Physico-Chimie Biologique, vol. 65, pp. 44–45.

Lewontin, Richard Charles (2001): *In the Beginning Was the Word*. In: Science, vol. 291, no. 5507, pp. 1263–1264.

Lilley, Irene M. (1967): *Introduction to Friedrich Froebel: a selection of his writings*. In: Lilley, Irene M. (ed. and tr.): Friedrich Froebel: a selection of his writings. Cambridge University Press, pp. 1–30.

Lister, David (1997): *Some Observations on the History of Paperfolding in Japan and the West – a Development in Parallel*. In: Miura, Koryo / Fuse, Tomoko / Kawasald, Toshikazu / Maekawa, Jun (eds.): Origami Science and Art: Proceedings of the Second International Meeting of origami Science and Scientific origami. Shiga: Seian University of Art and Design, pp. 511–524.

Liu, Lydia H. (2010): *The Freudian Robot: Digital Media and the Future of the Unconscious*. Chicago / London: University of Chicago Press.

Marín-Padilla, Miguel (2011): *The Human Brain: Prenatal Development and Structure*. Heidelberg: Springer.

Martin, Palmyre (1893): *L'Année préparatoire de travail manuel*. Paris: A Colin.

Miller, William Haig / Macaulay, James / Stevens, William (eds.) (1855): *The Kinder Garden*. In: The Leisure Hour: A Family Journal of instruction and recreation, vol. 204, pp. 743–745.

Mirsky, Alfred / Pauling, Linus (1936): *On the Structure of Native, Denatured, and Coagulated Proteins*. In: Proceedings of the National Academy of Sciences, vol. 22, no. 7, pp. 439–447.

Pasteur, Louis (1848): *Sur les relations qui peuvent exister entre la forme cristalline, la composition chimique et le sens de la polarisation rotatoire*. In: Annales de Chimie et de Physique, 3rd series, vol. 24, no. 6, pp. 442–459.

Prusinkiewicz, Przemysław / Barbier De Reuille, Pierre (2010): *Constraints of space in plant development*. In: Journal of Experimental Botany, vol. 61, no. 8, pp. 2117–2129.

Quastler, Henry (1958): *A Primer on Information Theory*. In: Yockey, Hubert P. (ed.): Symposium on information theory in biology. Gatlinburg, Tennessee, New York: Pergammon Press, pp. 3–49.

Richter, Fleur (1994): *Die Ästhetik Geometrischer Körper in der Renaissance*. Stuttgart: Gerd Hatje.

Row, Sundara Tandalam (1893): *Geometrical exercises in paper folding*. Madras: Addison Co.

Schäffner, Wolfgang (2007): *Electric Graphs: Charles Sandres Peirce und die Medien*. In: Franz, Michael / Schäffner, Wolfgang / Siegert, Bernhard / Stockhammer, Robert: Electric Laokoon: Zeichen und Medien, von der Lochkarte zur Grammatologie. Berlin: Akademie Verlag, pp. 313–326.

Schäffner, Wolfgang (2016): *Punkt 0.1. Zur Genese des analogen Codes in der Frühen Neuzeit*. Berlin / Zürich: Diaphanes.

Schmandt-Besserat, Denise (1992): *Before Writing, Vol. I: From Counting to Cuneiform*. Austin: University of Texas Press.

Schneider, Birgit (2007): *Textiles Prozessieren*. Zürich / Berlin: Diaphanes.

Schrödinger, Erwin (1944): *What is Life?: With Mind and Matter and Autobiographical Sketches*. Cambridge: Cambridge University Press.

Semper, Gottfried (1983): *London Lecture of November 11th, 1853: Outline for a system of comparative Style-Theory*. Ed. by Mallgrave, Harry Francis. In: RES Journal of Anthropology and Aesthetics, no. 6, pp. 8–22.

Shannon, E. Claude (1938): *A Symbolic Analysis of Relay and Switching Circuits*. In: Transactions of the American Institute of Electrical Engineers, vol. 57, no. 12, pp. 713–723.

Slater, Robert (1888): *Telegraphic code, to ensure secrecy in the transmission of telegrams*. London: W. R. Gray.

Soledad, Esteban (2008): *Liebig–Wöhler Controversy and the Concept of Isomerism*. In: Journal of Chemical Education, vol. 85, no. 9, pp. 1201–1203.

Thom, René (1975): *Structural Stability and Morphogenesis: An Outline of a General Theory of Models*. Trans. by Fowler, David H.. Reading et al.: W. A. Benjamin.

Turing, Alan (1936): *On computable numbers, with an application to the Entscheidungsproblem*. In: Proceedings of the London Mathematical Society, Series 2, vol. 42, pp. 230–265.

van 't Hoff, Jacobus H. (1877): *Die Lagerung der Atome im Raume*. Trans. by Herrmann, Felix. Braunschweig: Vieweg.

Wells, Kentwood D. (2007): *The Ecology and Behavior of Amphibians*. Chicago: The University of Chicago Press.

Yčas, Martynas (1958): *The Protein Text*. In: Yockey, Hubert P. (ed.): Symposium on information theory in biology. Gatlinburg / New York: Pergammon Press, pp. 70–102.

Michael Friedman; Wolfgang Schäffner

Email: *michael.friedman@hu-berlin.de*
Email: *schaeffner@culture.hu-berlin.de*

Image Knowledge Gestaltung. An Interdisciplinary Laboratory.
Cluster of Excellence Humboldt-Universität zu Berlin.
Sophienstrasse 22a, 10178 Berlin, Germany.

Karin Krauthausen

Folding the Narrative: The Dimensionality of Writing in French Structuralism (1966 – 1972)

Dimensionality

When in 1884 under the pseudonym A Square, Edwin A. Abbott wrote his novella *Flatland: A Romance of Many Dimensions* about the perception of two-dimensional beings, his principal concern was to expand the consciousness of the three-dimensional beings making up the book's readership.[1] Accordingly, these late-19th-century readers should be helped to visualize the possibility of a fourth dimension; and because in a world known as three-dimensional this fourth dimension was not so readily perceivable, an expansion of consciousness was required by means of the literary stimulation of the reader's imagination. In this way, Abbott's novella also helped the sciences and partic-ularly mathematics, since in the 19th century the latter had turned towards objects of knowledge of uncertain ontological status. A well-known example were the non-Euclidean geometries – that is, geometric systems that were internally consistent, but rested on different axioms than those of the geometry handed down since Euclid, which seemed to correspond so perfectly to the three-dimensional world, and thus in the 17th century became the ideal tool of the newly differentiating natural sciences, as well as, in *more geometrico,* a favored method of philosophical argumentation.[2]

---

1   Abbott/A Square 1884. In the second edition from the same year, 1884, the editor (presumably Abbott) in the *Preface to the Second and Revised Edition* addresses this reader in the name of his friend from Flatland, the fictive author A Square, in this sense: namely, as "readers and critics in Spaceland." Anonymous 1884, IX.

2   The argumentation *more geometrico* can be found in the work of René Descartes and Baruch de Spinoza among others. On the ontological question in the mathematics of the 19th century, see Gray 1992.

In the 19th century the construction of alternative geometries and the hypothesis of unknown dimensions shook up the sciences as well as the common knowledge of the period, and to compensate required an altered view of reality. The novella *Flatland* set out to provide such a compensation and, to this end, made use of an analogy. To the two-dimension characters of the novella, the Euclidean geometry based on a three-dimensional world would have to appear just as *fictitious* as the non-Euclidean geometries to many three-dimensional readers (including some mathematicians).[3] Abbott's novella corrects this judgment of fictitiousness and attributes to the new epistemological objects at least credibility and thus conditional reality. At the same time, he schooled the reader in the conviction that reality should, on the basis of science, be augmented by an invisible element, since at least the fourth dimension (as well as further dimensions and hence worlds) awaited discovery and axiomatic formulation. In the 19th century the world felt to be real was fraying both scientifically and narratively into the imaginary; in other words, reality seemed to contain further worlds hidden inside itself as if these were folded into being.

The embrace of a fourth dimension in the 19th century led to encounters not only between science and literature, but also between science and spiritualism, which despite their differences formed a generally binding consciousness of the dimensionality of the world.[4] Here, as an education in the new dimensions, literature acted as a welcome ally; however, it also represented a danger, since either it neglected epistemological analogies in favor of a science *fiction* (e.g. Jules Verne) or it *formalized* its own world by declaring speech and writing to be its basis in reality (e.g. Stéphane Mallarmé).[5] In neither case does it escape the awareness of dimensions, since also the literary and scientific consideration of speech and writing from the end of the 19th century onwards notes a dimensionality of these media. A dimension of writing and narration is explicitly determined by French structuralism and neostructuralism in the 1960s. Surprising with regard to the struc-

---

3   In the second half of the 19th century the analogy was ennobled by James Clerk Maxwell into a method for acquiring scientific knowledge. See Maxwell 1890, 155–159.

4   See Macho 2004; Henderson 1983.

5   On literary science fiction, see for example *Vingt mille lieues sous les mers* (*Twenty Thousand Leagues Under the Sea*, 1873) by Jules Verne (appearing in installments from March 1869 to June 1870; published as a book in 1871), a novel that in the activities of Captain Nemo allegorizes the 19th century passion for discovery and the hypostatized unknown worlds within the known world. The name of this figure, who operates as hinge between the worlds, recalls the figure of Odysseus in Homer's *Odyssey*, and thus that world-wanderer who in the episode with the Cyclops identifies himself verbally as Nobody, which leads to the paradoxical result that *Nobody* gouges out the eye of the Cyclops. For a *formalization* of the poetic literature of Stéphane Mallarmé and especially his late work, see *Un coup de dés jamais n'abolira le hasard* (1897), which activates as literary elements the sheet and the blank areas between the letters via typography and the layout of the page, and thus reflects on the material conditions of writing and literature as well as of reading. For a literature that makes use of methodological analogies to the sciences, see the cycle of novels *Les Rougon-Macquart* (1871–1893) by Émile Zola (with reference to heredity) and Paul Valéry's texts on Monsieur Teste, which attempt a physiological and psychological modeling of consciousness (first text: *La Soirée avec Monsieur Teste*, 1896).

turalist concept of dimension is that this is conceived rather simply, since, as in the 19th century, it denotes a direction of Euclidean space: that is, length (dimension 1), width (dimension 2), and height (dimension 3). This concept of dimension is expanded only insofar as it is now understood more abstractly as the degree of freedom of movement or of a particular position. Although in the early 20th century mathematics increasingly diversifies and complicates the concept of dimension (for example in Hermann Minkowski's concept of space-time from 1907–1908, the Hamel dimension in a vector space from 1905, and the so-called Lebesgue covering dimension or topological dimension of the 1930s), the general understanding of dimension remains relatively conventional, since it is still basically oriented to the three dimensions of Euclidean space. The fourth dimension in this general understanding is primarily identified – at the latest, since the popularization of Albert Einstein's special and general theory of relativity (published in 1905 and 1915 respectively) – with time, while both space and time are understood as relative values dependent on gravity (or matter). For Abbott and his contemporaries the fourth dimension in contrast was of a radically unknown nature, and as invisible as the third dimension for the two-dimensional beings in *Flatland*. This can be gathered from the remarks of the (fictitious) two-dimensional author A Square quoted in the preface . to the second edition by the anonymous editor:

> "It is true that we have really in Flatland a Third unrecognized Dimension called 'height,' just as it is also true that you have really in Spaceland a Fourth unrecognized Dimension, called by no name at present, but which I will call 'extra-height.' But we can no more take cognizance of our 'height' than you can of your 'extra-height.'"[6]

Hence, the experience or attribution of dimensions is not self-evident, but historically determined. In the following an episode of contentious dimensionality in structuralist analyses of writing and narration from 1966 to 1972 will be reconstructed and critically contextualized through references to the time-bound materiality of writing and narration in the book, and in this context to implicit and explicit "dimensional prejudices" (as Abbott would call them).[7] Here, the trickster position between the dimensions is assigned a material operation from book production: the fold.

---

6   Anonymous 1884, X. In the logic of the narrative and paratext, the author A Square is imprisoned in his homeland Flatland for asserting a third dimension, because he is not able to show either how this could be *measured* nor what *direction* it opens up. Accordingly, his cohabitants are *prejudiced* or lacking in *faith*. See ibid, XII.

7   Ibid.

Dimension 1

In 1966 no. 8 of the journal *Communications* appeared with contributions by among others the philosopher and semiologist Roland Barthes, the literary theorist Gérard Genette, and the philosopher and semiologist Tzvetan Todorov.[8] The issue focused on the theory of literary narrative, and the contributions by the above-named authors have in retrospect led to this issue of *Communications* being perceived as the "founding document of a 'French school' of structural analysis of narrative"[9] – whereby the "tentative profile"[10] set out in the issue was expanded on in later texts. In his contribution Barthes names structuralism as a framework and method when he separates narrative from its diverse "vehicles" (language, image, gesture, etc.) to explain the thus won *universal fact of narrative* via "an implicit system of units and rules."[11] Barthes is concerned with the hypothetical creation of a generally valid "structure of narrative," which he also describes as a "code," one that the narrator is aware of and able to activate.[12] For the planned scientific deduction of the narrative structure, Barthes draws (as do the structuralists in general) on the model of linguistics, since this attempts to come to grips with the variance of languages via the concept of the sign and, in the phonology of Roman Jakobson and Morris Halle, via the identification of a universal binary system of twelve distinctive features (*Fundamentals of Language*, 1956). The structuralists' understanding of structure is influenced by the concept of code found in information theory – similar to biology's integration of the concepts of code and program already from the 1940s to the 1960s. The promise of this methodological decision lies in the attainment of a universal and formal systematics that is able to reduce a multitude of isolated phenomena to a few elements and the rules for their combination. In her historical investigation *Who Wrote the Book of Life*, Lily E. Kay has shown in detail how after World War II biology adopted the concepts of information and code from information theory into biology, which in genetics became driving metaphors that were able to bring together different lines and approaches of research, and eventually contributed to the molecular biological deciphering of the genetic code.[13] Here, right from the beginning, the concept of code was linked to the concept of writing, since the genetic code is understood as a *transcription* – that is, the regulated transcription of a particular sequence of base pairs (i.e. genetic information) contained in DNA into a nucleotide sequence of the RNA strand, as well as the subsequent translation into the amino acid sequence of a protein. The analysis of this *genetic writing* hypostatized as

---

8   See Barthes 1966; Todorov 1966; Genette 1966. In the following, quoted from the English translations.

9   Vogt 1998, 300.

10   On the status of the contributions gathered in *Communications* no. 8 as a "tentative profile" for the searched-for structural theory of narrative, see Barthes 1975, 243.

11   Ibid, 237 and 238.

12   Ibid, 238, and for the equation of structure and code 238, footnote 2.

13   See chapters 2 and 3 in Kay 2000.

universal – thus, one that is principally the same in all known living beings – seemed to make it possible finally to decode the 'book of nature' (another, though earlier, metaphor).

Such ambitious ideas also took hold in French structuralism and especially the structural analysis of narrative in *Communications* no. 8. In his search for universal structures Barthes dissolved the boundaries of narrative both temporally and spatially, and he also *naturalized* it in order to subject it to the same fate experienced by the manifold phenomena of life in the concept of the genetic code: "narrative remains largely unconcerned with good or bad literature. Like life itself, it is there, international, transhistorical, transcultural."[14] The thus prepared *fact* of narrative should be reduced to a few elements and understood in analogy to the largest unit of linguistics, the sentence: "a narrative is a large sentence, just as any declarative sentence is, in a certain way, the outline of a little narrative."[15] Thus, only now, after the preparation of the object of knowledge by means of generalization and structural reduction, does Barthes identify elements and rules. While all the theorists in *Communications* refer to slightly different elements, all adhere fundamentally to the concept of a linguistic sign and to grammatical categories of the sentence and verb conjugation in linguistics. In Genette's case this orientation leads – in the elaboration of his narrative theory in *Discours du récit* (1972) – *en passant* to a rigid determination of the *direction* of the narrative and thereby to the allocation of a dimension. In its written form (here, Genette partly revokes Barthes's forgetting of the "vehicle") narrative thus belongs to an order of successivity; that is, it should be understood as a coded sequence of signs, and prescribes a diachronic reading. The dimension of written narrative is linearity (*dimension 1*), and its measure is found in the time of reading. Hence, narrative

> "can only be 'consumed,' and therefore actualized, in a *time* that is obviously reading time, and even if the sequentiality of its components can be undermined by a capricious, repetitive, or selective reading, that undermining nonetheless stops short of perfect analexia: one can run a film backwards, image by image, but one cannot read a text backwards, letter by letter, or even word by word, or even sentence by sentence, without its ceasing to be a text. Books are a little more constrained than people sometimes say they are by the celebrated *linearity* of the linguistic signifier, which is easier to deny in theory than eliminate in fact. [...] [P]roduced in time, like everything else, written narrative exists in space and as space, and the time needed

---

14 Barthes 1975, 237.
15 Ibid, 241. Literature is equated by Barthes with language and only as a consequence of this equation can linguistic analysis serve as a model for a universally valid analysis of narrative. Fundamental for such an analysis of narrative oriented to linguistics is then the code, which means that in the narrative nothing is accidental, but everything coded and therefore explainable via structure or organization.

for 'consuming' it is the time needed for *crossing* or *traversing* it, like a road or a field."[16]

Thus, for Genette written narrative exists in three-dimensional space, but is itself strictly one-dimensional, and precisely in this way available to a structural analysis, insofar as the stream of text (a string of linguistic signs) relates to universal laws of language. Such equations of writing and linearity, which are typical for structuralism, were criticized by Jacques Derrida in *De la grammatologie* (*Of Grammatology*) already in 1967 (thus, a year after the papers in *Communications* no. 8 and five years before Genette's elaboration of his narrative theory). Derrida's neostructuralist position introduces historical relativizations and philosophical subversions into the structuralist analysis of speech and writing.[17] In his project of a liberated "science of writing – *grammatology*" (writing understood in a general sense as inscriptions of all kind, and thus as *grammè* or graphs), he traces the "false evidence" that determines the long history of writing.[18] This includes particularly the orientation of writing to spoken language and a model of presence, since this orientation leads to a focusing on a *phonetic writing* and overlooks other forms of writing (ideographic writing, knot writing, etc.), or entails the degradation of cultures "said to be 'without writing.'"[19] Also the "production of the linear norm" (for which Genette is responsible in an exemplary way, but not alone) is for Derrida the expression of an approximately four-thousand-year-old ideology that should be corrected in the proposed science of grammatology via a liberated and radicalized concept of writing by among other things invoking the pre- and parallel histories of nonlinear writing.[20] Derrida's critique outlines something that could be described with Abbott's words as "dimensional prejudices" – namely, an absolutized linearity of writing (and its unreflected relation to a conventional concept of temporality). The supposed universality of linear writing is countered by Derrida with its historical relativity: "The 'line' represents only a particular model, whatever might be its privilege. This model has *become* a model and, as a model

---

16 See Genette 1980, 34 (unless otherwise noted, emphasis in the original). Here, Barthes is more cautious and combines the analysis of the syntactic order of the sentence with the consideration of a paradigmatic order (in the sense of linguistics: syntagmatics and paradigmatics), which undermines the linearity of the stream of text in the narrative. See Barthes 1975. For the usual equation of writing and alphabetic writing and finally linearity, see a recent publication on the history of writing and the book, Funke 1992, 13–28.

17 Structuralism (e.g. Todorov, Genette, Barthes) and neostructuralism (e.g. Derrida, Michel Foucault) do not only overlap temporally, but are also to be understood content-wise as networked and complementary positions – the term poststructuralism masks this connection. See Dosse 1991–1992.

18 Derrida 1976, 4 and 81.

19 See ibid, 83: "Actually, the peoples said to be 'without writing' lack only a certain type of writing. To refuse the name of writing to this or that technique of consignment is the 'ethnocentrism that best defines the prescientific vision of man.'"

20 One of the representatives of this writing ideology criticized by Derrida is the linguist Ferdinand Saussure (see ibid, 86).

it remains inaccessible."[21] In the course of his deconstruction of phonetic writing, he not only reconstructs the history of this ideology, but also recalls those elements that constantly endangered the linear norm, including the "discreteness" and "spacing" resulting from the respective storage media.[22] These moments of crisis of linear writing should now – beyond Derrida – be explicitly taken up again by describing the material condition of writing (and narration) in the book: that is, the two- and three-dimensionality of writing in codex and bound book neglected by structuralism.

Dimensions 2 and 3

If one wants to illustrate the hypostatization of linear writing criticized by Derrida with a book medium, then the bound book produced since the 15th century is not strictly speaking the best choice. It seems more appropriate to illustrate linearity with media such as Egyptian, Greek, and Roman scrolls, which – after the appropriate material (papyrus or leather) had been found – with their maximum length of roughly six meters (a book format of the time; today more like a chapter) permitted the writing down of corresponding text lengths, and due to their rolling technique prescribed a relatively clear reading direction. Other material supports such as clay shards or tablets made possible mostly only shorter entries, or had to be provided with short inscriptions along the lower edge (colophons) to indicate their relation to a larger set of tablets. The codex developed out of the wooden (or wax) tablets that already could be bound with leather straps to form diptycha, triptycha, or polyptycha, so that the writing material (papyrus, parchment, and, after the 11th century when it was imported to Europe, also paper) was no longer glued together to form a roll, but folded after the model of polyptycha to form a *section*, so that subsequently a number of sections could be bound together. That increased manageability by making it possible to include longer manuscript texts and entries in one codex.[23] The beginning of the use of paper in western Europe in the 11th century, the production of paper beginning in the 13th century, and finally, beginning in the second half of the 15th century, the invention of printing with type, which with the invention of new type-casting methods became easier to produce, led to a considerable acceleration in book production. According to current research on the history of the book, for the second half of the 15th century one can suppose a total production of approximately 15 million

---

21 Ibid. On the binding of the linearity of language and writing to "this vulgar and mundane concept of temporality (homogeneous, dominated by the form of the now and the ideal of continuous movement, straight or circular)," see ibid. This concept of time rejected by Derrida is at the basis of Genette's remarks on the temporality of reading.

22 Ibid.

23 On the history of the clay tablet, scroll, and codex, see Funke 1992, 66–71. On the success of the codex, see ibid, 70. Editions of the Bible are known in codex form already in the 1st century; editions of Homer and Virgil date from the end of the 1st century; from the 4th century onwards the codex gradually replaced the scroll.

books – however, mass production in the modern sense only arose with the invention of paper machines, folding machines, typesetting machines, industrial type-casting methods, and high-speed presses at the beginning of the 19th century.[24]

As a storage medium for longer texts the codex does not immediately align itself with the scroll. On the one hand, the continuous format of the scroll is now divided up into square and, later, rectangular pages, which present a specific quantity of text as in a frame. This type of discrete page is partly anticipated in the scrolls to the extent that here the text is entered in columns whose line length is initially oriented to the form of the Greco-Roman hexameter, so that not only each column, but already each line can be understood as a verse, and in this way as a discrete unit (to which always the next line or column is linked). However, the single page in the codex is not oriented to a metric form, but determined only by the storage medium itself, which interrupts the continuity of the text. Furthermore, the length of the texts in a codex can be much longer than in a scroll, so that the connection between the pages in some circumstance has to continue over a considerable length of text. Thus, while the two-dimensional presentation of text in the scroll with its line and column breaks can be seen as a minor disruption of the linear stream of text, for the codex the two-dimensional presentation of the text on the page is grasped more discretely; and in addition through the folding of the paper and the binding of the resulting sections a third dimension becomes noticeable that decisively alters what can be treated in writing (narratives, laws, science, history, etc.) by grasping this spatially. The discrete pages and the spatial sections of codex and book prevent a strict linearity of the text; or formulated against the background of the structuralist discussion on dimensions, in codex and book the one-dimensionality of writing is supplemented by two- and three-dimensionality. This increased dimensionality of the book medium is made possible by a practice that initially remains external to the text: the folding of the sheet of paper. While wooden tablets have to be bound together using leather straps and thus simply added, the wondrous proliferation of the textual space in codex and book is the result of a multiplication: that is, a single or repeated folding of the sheet of paper to create formats of different sizes (customary is a threefold folding of the sheet to produce 16 pages; see fig. 1). The folded sections can then be combined using different binding techniques (sewing and/or gluing) to form extensive codices or books, and are subsequently trimmed at the edges so that the continuous fold of the paper at the unbound but folded outer edges is interrupted to create a sequence of pages that can now be leafed through. Folding and trimming thus bring about a two- and three-dimensionality of the text in codex and book, and such a bound collection of sheets finally became a full totality when after the mid-15th-century printing facilitated and even provoked the creation of longer texts; now long textual narratives became quasi-mechanically their own worlds.

---

24   See ibid, 113, and on industrialization 189–199.

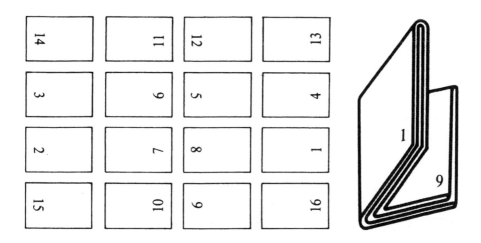

Fig. 1: A current example of a 16-page imposed forme. After being folded in the folding machine, the pages are arranged in the correct sequence. (Drawings by Bernhard Laufer).

In the historical vocabulary of the book manufacturer the fundamental significance of the transition to *dimension 3* produced by the folding of the sheet of paper has been retained to the present day. In the German-speaking world the proper arrangement of the pages on a sheet of paper, which ensures that the correct sequence appears in the book after folding, is called *ausschiessen* (imposition). In the etymological volume of the *Deutsches Wörterbuch von Jacob und Wilhelm Grimm* one can read that the verb has several meanings, whereby the transitive "einen Wald ausschieszen" (to shoot all wildlife in a forest) surely comes closest to the economical utilization of the sheet (with no leftover space).[25] Nevertheless, one should also consult an intransitive usage, since *ausschiessen* in the sense of *prosilire, exsilire,* and *progerminare* (and still more frequently *schieszen* [to shoot]) is also used in relation to plants, and signifies for instance the burst of growth in spring, when "die bäume ausschieszen" (the trees put out their shoots).[26] Accordingly, there is an abrupt change in which the plant, suddenly and profusely, puts out new shoots. This remarkable plant growth can be compared with the folding of the paper sheet, since in effect the threefold fold of a single large sheet after trimming leads in one stroke to a 16-page section – the book could be said to burgeon.[27]

---

25 Grimm 1991, vol. 1, cols. 948–949, here col. 949.

26 Ibid.

27 Drawing on Barthes's formulation that narrative like life is simply "there" (cf. above in the main text), one could say – with regard to the current state of knowledge on the fundamental importance of protein folding for the development of living organisms – that in the eyes of book manufacturers the pages of the folded and trimmed book are suddenly there, like life.

Also the German term for folding, *Faltung* (or *Falzung*), which book manufacturers employ for this operation, contains revealing historical connotations. Thus, for bookbinders the technical term *Falz* (fold) means historically "das brechen, zusammenlegen, falten der gedruckten bogen" (the breaking, arranging, and folding of the printed sheet), and thus includes not only the folding and assembling, but also the "breaking," which means the term implies a discontinuity.[28] As a result of the assembling, the *Faltung* also creates a split, which is why the word *falte* (fold) historically also includes the old meaning of an "abgelegenen, eingehegten raums" (remote, enclosed space).[29] These historical meaning variants of *ausschiessen* and *falte* already shed light on the new praxis of reading demanded by the extended textual space of codex and book. Since, in contrast to the scroll, the bound book allows and provokes a leafing back and forth, and thus a looking ahead and back, or a random perusal of the book. Each time one turns the page in the continuous sequence of pages (shown concretely by the page numbers), what was hidden now emerges in the space between the sheets, and can be read on the opened page. At the same time, each turning of the page can be deployed and understood as a break, since from here one can leaf to any other part of the book, and thereby create a new relation. Thus, the successivity of the narrative emphasized by Genette is replaced in the practice of reading by a network-like relation between different parts of the narrative, since a narrative stored in a book can indeed be read backwards or in a crisscross fashion or even only in extracts.[30] In practice the discontinuous reading suspends the alleged linearity of the text, and it can do this with such ease because the material organization of the book (discrete sheets or pages) enables this. Although the fold in the paper first brings about the book as totality (in the sense of a whole world), the same fold leads the reader to fragment this totality through reading and to disrupt the predetermined order.

Book manufacturers were aware that the folding and binding of the sections did not only mean continuity, multiplication, and totality, but also contained the danger of discontinuity and confusion. The problem regarding the imposition (*ausschiessen*) of the sheet of paper was then also to ensure the proper arrangement of the pages to be printed, so that in the folded, bound, and trimmed book the pages appeared in the right sequence (and also that the grain of the paper was parallel to the spine making the pages easier to turn).[31] Today, this arrangement is carried out by an algorithm; however, up to the 20th century it was common practice to insert the pages into a chase divided according to the

---

28 Grimm 1991, vol. 3, cols. 1302–1303, here col. 1303.
29 Ibid, cols. 1297–1299, here col. 1298. Other meanings of *falte* are *umschlingung* (intertwining), *gefaches* (dividers/compartments), and *tasche* (pocket), and naturally the *falte* in a garment, or in the figurative sense "falte des herzens, der sinne." (fold of the heart, the senses). Ibid.
30 At least if one does not take the string of characters as an absolute criterion (as Genette does), but starts from units such as page, chapter, and paragraph, while also considering reading and writing practices.
31 On the problems of the division of the page and folding, see Diderot/d'Alembert 1751–1780, vol. 29, Imprimerie en caractères, Planches VI–XII (and the descriptions in ibid., 3–7); Febvre/Martin 1979, 68–71; in relation to the present, Laufer 1988, 65–71.

number of pages and folds (see fig. 2), and to make a trial pull, or else to print the pages individually, to paste them onto a sheet of paper, and to fold this to make a test copy. Without making such tests, the result of the folding would be uncertain; which means that with a defective arrangement after printing and folding, while the linearity of the text is formally preserved (as a mere string of characters), the meaning of the text is disrupted.

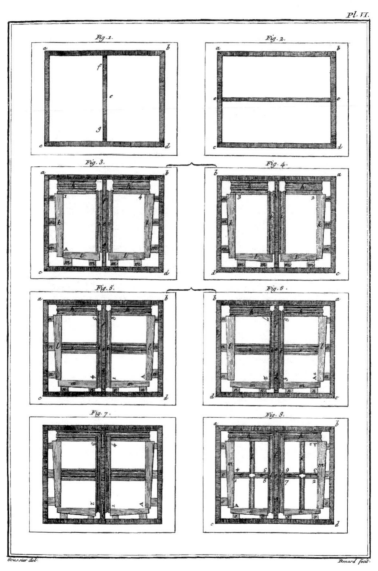

Imprimerie, Impositions .

Fig. 2: The illustration from the plates section of the *Encyclopédie* of Diderot and d'Alembert shows a few simple examples of the numerous possible divisions of the page in the c hase. In the 18th century different divisions were selected depending on the type of paper, folding, and binding.

Of course, the danger of discontinuity linked to the fold can also be made productive. In the 1960s the American writers William S. Burroughs and Brion Gysin took printed books (by other authors), folded the pages down the middle, and arranged the two halves of the page in such a way that a new page with a now hybrid text resulted. This creative writing method propagates a *folding of folding* that at first glance retains the linearity of the text, but de facto disrupts and transforms the cohesiveness of the text.[32]

Fundamental for the printed, folded, and bound book from the 15th to the 21st century is therefore not the one-dimensional text, but the spatializing folding, which variably links continuity and discontinuity, and from the *splitting* of the sheets (i.e. the trimming of the fold) allows a world to arise. In this formulation of the book and the corresponding reading techniques, the text is at least three-dimensional.

Many Dimensions

In *De la grammatologie* Derrida declared the four-thousand-year-old ideology of linear text to be concluded, and in this he included the book: "The end of linear writing is indeed the end of the book, even if, even today, it is within the form of the book that new writings – literary or theoretical – allow themselves to be, for better or for worse, encased."[33] For his diagnosis of the end Derrida draws on the investigations of André Leroi-Gourhan, who in 1964 and 1965 published the volumes of *Le geste et la parole* (*Gesture and Speech*, 1993) on language and operativity. Here, the anthropologist recalls those epochs of writing in which writing was not organized according to a linear schema, but operated in a multidimensional way, and thereby constituted a correspondingly multidimensional thought. At the same time, however, Leroi-Gourhan also announces a new development, which he describes as a liberating "return to diffuse and multidimensional thought."[34]

---

32 A comparatively conventional result of this experimental project is the publication *The Third Mind* by both authors from 1978.

33 Derrida 1976, 86. Twenty years later he would revise this estimation in *Le Livre à venir* (*The Book to Come*; an introduction to a discussion with Roger Chartier and Bernard Stiegler that took place in 1997 at the BnF in Paris), where he takes up again the question of the book and here ascribes to it a long life, one that is, however, subjected to considerable transformations through the material, technical, and social practices associated with it. The revolution of the book that he considers here is digitalization, which exposes the book to a changed support and changed conditions. As a result, especially the book as *totality* is abandoned (instead, tension between "gathering" and "dispersion") as is the finalized book (the folded and printed book effectively stops and substitutes the writing process of the author). However, he already sees this model of the book announced in Mallarmé's *Un coup de dés*. See Derrida 2005, 13. In this book Derrida only briefly comes back to his remarks in *De la grammatologie*. In his earlier diagnosis of the *end of the book*, he sees the discussion on the book in the digital age already implied. With end, however, he means in retrospect simply the *end of the codex-book*. See ibid, 14f.

34 André Leroi-Gourhan, *Le geste et la parole* (1965, vol. 2, 261–262), here quoted in Derrida 1976, 332f., footnote 35.

And Derrida confirms the significance of this turn that should end the "age of the sign" and linearity: "What is thought today cannot be written according to the line and the book, except by imitating the operation implicit in teaching modern mathematics with an abacus."[35]

Surprisingly, Leroi-Gourhan and Derrida's diagnosis was sparked by two phenomena that could also be claimed by the proponents of a linearity of writing. For the anthropologist it is magnetic tape and the Dictaphone that has put an end to the dominance of the linearly organized textual space in the book; it is therefore a storage of spoken language (which served as the basis for the phonetic writing and presence model criticized by Derrida) that is evoked here as the return of many dimensions. Derrida, on the other hand, refers to the phenomenon of *program* or *genetic inscription,* since here writing has no *anthropos* and no intentional consciousness as a *source* and point of reference, but can vouch for the radicalized understanding of writing outlined in *De la grammatologie.* Accordingly, the genetic inscription attests to a "history of the *grammè*" in which the possibility of human writing as human unity in general will have been a mere "stage or an articulation in the history of life."[36] To the extent, however, that particularly genetic inscription might be understood also as evidence of a universality of the writing paradigm, it is still possible to detect with regard to Derrida an interface to the ideology of linear writing.

However, a three-dimensional writing (and a corresponding thought) can also be observed before the invention of magnetic tape and Dictaphone, and this applies beyond the practice of reading. The printed and folded book is accompanied by another writing practice insofar as the finalized book is eventually chopped into parts again through reading and excerpting, and these parts can be recombined by means of writing to form a new entity. Decisive for this is the invention of the card index, which encourages this reading and writing activity. In the 16th century scholars such as the polymath Konrad Gessner attempted to filter their reading by excerpting the text on small pieces of paper – the precursors of index cards – and in this way to channel these excerpts in a new direction from a particular perspective. Thus, through reading, the totality of the book is broken into pieces, the searched for passages recorded on discrete paper formats (excerpted), and then in further acts of reading and writing (in some cases initially in the form of additional notes, also on index cards) combined to form new connections from which a new book might arise. The three-dimensional card index (folders, cases, boxes, or whole cabinets with draws for the cards) brings these notes into a constellation that can be differently organized (according to area of knowledge, the subject, or idiosyncratic perspectives)

---

35 Derrida 1976, 87. On the "age of the sign" as an ideology, see ibid, 14. See also Barthes's cautious relativization of the structuralist claim to universality, when in his text *Sémiologie et Médecine* from 1972 he likewise asks about a possible "ideology of the sign." See Barthes 1988, 212. Historical relativity and the diachronic level returned to structuralism at the beginning of the 1970s.

36 Derrida 1976, 84.

and thereby from book reading to book reading gradually expanded and altered. Unlike in a book, however, index cards are not placed in a fixed and closed succession, but like playing cards retain a fundamental mobility (they can be repositioned and expanded) as well as the potential of recombination (insofar as the individual cards can at any time be recombined to form new relations).[37]

While in his experiments with systems of note taking, Gessner basically still follows the idea of a topical, sorted, and closed order of knowledge, one that is not oriented to originality and innovation, since the modern period in philosophy, the sciences, and the arts, an awareness formed for the eventful, human-made *new* which leads to the desire for its repetition.[38] This search for innovation is reflected in all areas of culture and leads eventually to a radical temporalizing of scientific, philosophical, artistic, and technological production – the horizon of knowledge is *open*. Under these conditions the card index developed into a favored means of administrating the increasingly vast knowledge of the accelerated book production and to create the sought-after new knowledge. At the same time, the practice of note taking is gradually differentiated into a diversely characterized, but also increasingly theorized method, so that at the beginning of the 20th century it was propagated as a universal technique, and the production of card indexes could be standardized.[39] The card-index method now fulfilled administrative tasks in bureaucracies, accounting departments, and libraries, but also characterized the writing processes of scientists and artists (see fig. 3), before being replaced everywhere by computers and the possibilities offered by these.[40]

---

37 See Krajewski 2011; Zedelmaier 2002.
38 On this change, see Blumenberg 1996. On the relation between card index file and evolution of knowledge, see Kammer 2009.
39 Here, writing with the help of the card index follows the method of scholarly excerpting (selection, summary, sorting, and storage), which it effectively radicalizes. On these principles of scholarly note taking, cf. Blair 2004. Note taking is initially subordinated to excerpting and theorized and professionalized as an independent technique only late. In the 18th century the technique becomes evident among other places in the descriptions of the jurist Johann Jacob Moser, to which the writer Jean Paul refers in his own practice of note taking. There are also early warnings of the dangers of the card index, for instance that this form of note taking (*Verzettelung*) leads to a getting lost in the details (*sich verzetteln*), and thereby no longer leads to the production of cohesive texts.
40 Among *scholarly* users of card index files of the 20th century one could name Roland Barthes, Michel Foucault, Hans Blumenberg, Niklas Luhmann, Arno Schmidt, and many others. For illustrative examples, see the exhibition catalogue of Gfrereis/Strittmatter 2013. On the order and use of the card index in the writing practices of Blumenberg, Luhmann, and Schmidt, as well as the related book production, see Krauthausen 2013.

Fig. 3: The photograph by Friedrich Forssman shows part of the vast card file (approx. 120,000 small pieces of paper) that the writer Arno Schmidt created in the 1960s for his novel *Zettel's Traum*.

The meaning of this writing practice, which accompanies the printed and folded book, was elucidated in 1928 by Walter Benjamin, when he notes that the "card index marks the conquest of three-dimensional writing."[41] The philosopher's remark makes reference to the spatial dimensions of the furniture that the card index concretely is, and that conditions it both materially and as medium, insofar as the mobile three-dimensional correlation of the index cards expands the dimensionality of writing.[42] What counteracts the linearity of writing in this case is not only the caesura brought about by a new medium, but a changed practice of reading and writing, which in turn is made possible by the changed material disposition of the book.

What, however, can be concluded from this description of the many dimensions of writing in the book and card index? Basically, a relation to the book as set out by Derrida against the background of his diagnosis of the end of the book:

---

41 Benjamin 1979, 62. Benjamin refers explicitly to the fact that at other times and in other cultures three-dimensional textual media had already existed, for instance in runes and knot writing.

42 If one considers the mobility of the cards and the open-endedness of the card index, then one should also add the temporal dimension. Accordingly, the card index would establish a four-dimensional writing.

"It is less a question of confiding new writings to the envelope of a book than of finally reading what wrote itself between the lines in the volumes. That is why, beginning to write without the line, one begins also to reread past writing according to a different organization of space."[43]

In this sense, it is praxis that breaks open the (structuralist) ideology of the linearity of writing. And it should be added that, strictly speaking, this had long been carried out by praxis: on the one hand, insofar as the practice of folding the paper sheet contained the possibility of many dimensions of writing; on the other, insofar as the accompanying practice of a discontinuous and spatializing reading (or such a writing) realized a potential inherent to the material conditions.[44]

Translated by Ben Carter

---

43 Derrida 1976, 86.
44 See in this connection also Barthes's considerations on the meaning of praxis, since this escapes the endless chains of signifiers of (clinical) signs and establishes a meaning through simple hermeneutics, which then determines the further operating. Barthes 1988, 210.

Bibliography

Abbott, Edwin A. / A Square (1884): *Flatland: A Romance of Many Dimensions*. London: Seeley & Co.

Anonymous (1884): *Preface to the Second and Revised Edition*. In: Abbott, Edwin A. / A Square: Flatland: A Romance of Many Dimensions. 2nd ed. London: R. Clay, Sons and Taylor, pp. IX–XIV.

Barthes, Roland (1966): *Introduction à l'analyse structurale des récits*. In: Communications, vol. 8, no. 1, pp. 1–27.

Barthes, Roland (1975): *An Introduction to the Structural Analysis of Narrative*. Trans. by Duisit, Lionel. In: New Literary History, vol. 6, no. 2, On Narrative and Narratives, pp. 237–272.

Barthes, Roland (1988): *Semiology and Medicine*. In: id.: The Semiotic Challenge. New York: Hill and Wang, pp. 202–213 (French original: Barthes, Roland [1972]: *Sémiologie et médecine*. In: Bastide, Roger: Les sciences de la folie. Paris: Mouton, pp. 37–45).

Benjamin, Walter (1979): *One-Way Street*. In: id.: One-Way Street and Other Writings. Trans. by Jephcott, Edmund / Shorter, Kingsley. London: NLB, pp. 45–104.

Blair, Ann (2004): *Note Taking as an Art of Transmission*. In: Critical Inquiry, vol. 31, no. 1, pp. 85–107.

Blumenberg, Hans (1996): *"Nachahmung der Natur": Zur Vorgeschichte des schöpferischen Menschen*. In: id.: Wirklichkeiten in denen wir leben. Stuttgart: Reclam Verlag, pp. 55–103.

Derrida, Jacques (1976): *Of Grammatology*. Trans. by Spivak, Gayatri Chakravorty. Baltimore / London: Johns Hopkins University Press (French original: Derrida, Jacques [1967]: *De la grammatologie*. Paris: Les Éditions de Minuit).

Derrida, Jacques (2005): *The Book to Come*. In: id.: Paper Machine. Trans. by Bowlby, Rachel. Stanford, CA: Stanford University Press, pp. 4–18.

Diderot, Denis / d'Alembert, Jean-Baptiste le Rond (1751–1780): *L'Encyclopédie: Suite du Recueil de Planches sur les Sciences, les Arts libéraux et les Arts mécaniques, avec leur explication*. Paris: Briasson et al.

Dosse, François (1991–1992): *Histoire du structuralisme*. 2 vols. Paris: La Découverte.

Febvre, Lucien / Martin, Henri-Jean (1979): *The Coming of the Book: The Impact of Printing 1450–1800*. Trans. by Gerard, David. 2nd ed. London: NLB.

Funke, Fritz (1992): *Buchkunde: Ein Überblick über die Geschichte des Buches*. 5th rev. ed. Munich et al.: K.G. Saur Verlag.

Genette, Gérard (1966): *Frontières du récit*. In: Communications, vol. 8, no. 1, pp. 152–163.

Genette, Gérard (1980): *Narrative Discourse: An Essay in Method*. Trans. by Lewin, Jane E. Ithaca / New York: Cornell University Press (French original: Genette, Gérard [1972]: *Discours du récit*. In: id.: Figures III. Paris: Les Éditions du Seuil, pp. 67–273).

Gfrereis, Heike / Strittmatter, Ellen (eds.) (2013): *Zettelkästen: Maschinen der Phantasie*. Exh. cat. Literaturmuseum der Moderne, Marbach am Neckar. Marbach am Neckar: Deutsche Schillergesellschaft.

Gray, Jeremy (1992): *The 19th century Revolution in Mathematical Ontology*. In: Gillies, Donald (ed.): Revolutions in Mathematics. Oxford: Oxford University Press, pp. 226–248.

Grimm, Jacob / Grimm, Wilhelm (1991): *Deutsches Wörterbuch von Jacob und Wilhelm Grimm*. 16 vols. Munich: Deutscher Taschenbuch Verlag.

Henderson, Linda Dalrymple (1983): *The Fourth Dimension and Non-Euclidean Geometry in Modern Art*. Princeton, New Jersey: Princeton University Press.

Kammer, Stephan (2009): *Zettelkasten und bewegliche Lettern: Die poetologische Entzauberung des Anfang(en)s*. In: Thüring, Hubert et al. (eds.): Anfangen zu schreiben: Ein kardinales Moment von Textgenese und Schreibprozeß im literarischen Archiv des 20. Jahrhunderts. Munich: Fink Verlag 2009, pp. 29–42.

Kay, Lily E. (2000): *Who Wrote the Book of Life: A History of the Genetic Code*. Stanford, CA: Stanford University Press.

Krajewski, Markus (2011): *Paper Machines: About Cards & Catalogs 1548–1920*. Trans. by Krapp, Peter. Cambridge, MA / London: The MIT Press.

Krauthausen, Karin (2013): *Supports entre ordre et désordre: Réflexions sur les fichiers de Hans Blumenberg, Niklas Luhmann et Arno Schmidt*. In: Genesis, no. 37, pp. 113–126.

Laufer, Bernhard (1988): *Basiswissen Satz Druck Papier*. 2nd rev. ed. Düsseldorf: Buchhändler heute / Triltsch Druck und Verlag.

Macho, Thomas (2004): *Das Rätsel der vierten Dimension*. In: Macho, Thomas / Wunschel, Annette (eds.): Science & Fiction: Über Gedankenexperimente in Wissenschaft, Philosophie und Literatur. Frankfurt a. M.: Fischer Verlag, pp. 62–77.

Maxwell, James Clerk (1890): *On Faraday's Lines of Force*. In: Id.: The Scientific Papers of James Clerk Maxwell, vol. 1. Ed. by Niven, William Davidson. Cambridge: Cambridge University Press, pp. 155–229.

Todorov, Tzvetan (1966): *Les catégories du récit littéraires*. In: Communications, vol. 8, no. 1, pp. 125–151.

Vogt, Jochen (1998): *Nachwort des Herausgebers*. In: Genette, Gérard: Die Erzählung. 2nd ed. Munich: W. Fink Verlag, pp. 299–303.

Zedelmaier, Helmut (2002): *Buch. Exzerpt. Zettelschrank. Zettelkasten*. In: Pompe, Hedwig / Scholz, Leander (eds.): Archivprozesse: Die Kommunikation der Aufbewahrung. Cologne: DuMont Verlag, pp. 38–53.

Karin Krauthausen
Image Knowledge Gestaltung. An Interdisciplinary Laboratory.
Cluster of Excellence Humboldt-Universität zu Berlin.
Sophienstrasse 22a, 10178 Berlin, Germany.

Angelika Seppi

Simply Complicated: Thinking in Folds

I) "Trying to see the grass in things and words"

"Those things which occur to me, occur to me not from the root up but rather only from somewhere about their middle", Franz Kafka writes in his *Diaries* and continues: "Let someone then attempt to seize them, let someone attempt to seize a blade of grass and hold fast to it when it begins to grow only from the middle."[1] In *A Thousand Plateaus* Gilles Deleuze and Félix Guattari pick up the thread and confirm that it is "not easy to see things in the middle, rather than looking down on them from above or up at them from below, or from left to right or right to left: [...]. It is not easy," they write, "to see the grass in things and in words [...]."[2] Trying to see the grass in things and words, to perceive of things and words in their becoming, is one of the many devices Deleuze and Guattari adopt to designate the main direction of their shared philosophical endeavor. Their multifaceted philosophical journey unfolds throughout a lifetime of thinking and writing, relentlessly stepping into new and uncertain terrain. All the elements constituting the open totality of their work resonate in an overall pursuit of an affirmative theory of multiplicity, difference and becoming. While multiplicity, difference and becoming are anything but novel philosophical concepts, it is their unconditional affirmation, which constitutes a departure from what has been labeled, since Friedrich Nietzsche, the underlying negativity/nihilism, which informs western thought from Socrates to Hegel and beyond. From Plato onwards – that is how the thread of European philosophy is traced by Nietzsche and weaved further by Deleuze and Guattari among others – the triumph of thought over life is synonymous with the triumph of the intelligible over the

---

1   Kafka 1948, 12.
2   Deleuze/Guattari 1987, 44.

sensorial, of the ideal over the material, of the one over the multiple, of identity over difference, of being over becoming.[3]

The paper develops the fold as a counter-figure and a counter-concept to any such dichotomous overlay. Taking as its point of departure Jacques Derrida's notion of the *undecidable* that resists and disorganizes philosophy's binary order from the very inside of the philosophical text itself, the paper pursues the fold as an exemplary figure of the *logic of the supplement* accordingly developed. By doing so the fold is further traced in its etymological links, its conceptual lineage and its material manifestations, and along the aesthetic riddles it poses. The abundance of terms, such as *simplicity, complexity, implication, explication, application, multiplication* – all of whom etymological derivations of the latin *plicare* (to fold), *plectere* (to plait, twine), following the greek *plékein* (to plait, to weave) – serves as an initial indicator to the degree to which our language and thought are permeated by folds and processes of folding. Regarding their material manifestations, they stretch out from the folds of drapery to the folds of living tissue, from the diptychs of antique tablets and reliefs to the explicit or implicit diptychs of painting, from book-folds to present-day folded Note-Books, from the art of folding paper to foldable architecture, from biological processes such as invagination or protein-folding to René Thom's famous morphological catastrophes. Correspondingly broad is the span of disciplines, within and beyond their limits – the fold extends itself, from philosophy to mathematics, biology, physics and chemistry, from the arts to art history, and so on.

Touching upon a variety of viewpoints, Gilles Deleuze's work *The fold. Leibniz and the Baroque* will figure as a thread weaved through the arguments developed in this paper. The guiding image will be provided by the Leibnizian-Deleuzian allegory of the Baroque house of thought – an allegory of the single, virtual plane that unfolds both the pleats of matter and the folds of the soul. (fig. 1) According to Deleuze, the world in general – encompassing the virtual plane that is unfolded through the pleats of matter and the folds of the soul – thus becomes comparable to an infinitely folded curve that extends to infinity.[4] Regarding the fold's complexity, the question arises, whether and how a comprehensive concept of the fold is possible at all. To develop a philosophical concept of the fold is certainly what Deleuze attempts to do in his reading of Leibniz. To retrace Deleuze's attempt to conceptualize the fold, will thus form the main focus of this paper. Finally, thinking in folds will be evoked as an attempt to re-conceptualize the distributions that constitute our world from the point of view of their becoming. With Deleuze and Derrida

---

3   A generalization such as this is merely provocative, of course, so that if ever there was any truth to the famous quip by Alfred North Whitehead, maintaining western philosophy to be no more than a series of footnotes to Plato (Whitehead 1978, 39), this series would have to comprise the countless attempts, dating as far back as Plato himself, to counter Platonism and its underlying dichotomous structures.

4   Cf. Deleuze 1993a, 24.

I will conclude the arguments presented in this paper by indicating an essential non-being and not-being-now that subverts the commonly assumed positivity and presence of being.

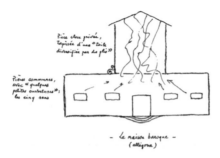

Fig. 1: Gilles Deleuze, The Baroque house (allegory).

II) Where does a fold begin and where does it end?

In his attempt to further advance a critique of western metaphysics Derrida pointed out not only to the hierarchical and dichotomous structure informing western metaphysics – presence, truth, identity or unity being prioritized over absence, error, difference or multiplicity – but also to the ambivalences lying at its core. Let us take, as an example, Derrida's famous reading of Plato's *Phaedrus*[5]. The dialogue between Socrates and Phaedrus moves from an initial query concerning love, to a discussion on the merits of speech in contrast to writing. It comes as no surprise that Socrates – digressing into the myth of Thoth, who figures as the inventor of the so called pharmakon of writing – condemns writing while positing direct speech as the only proper vehicle of truth. To be sure, Derrida is not concerned with presenting yet another evidence of western logo-centrism, but rather with unfolding a complexity intrinsic to the Platonic text itself. He points to a double meaning embedded in the text – *pharmakon* is both a remedy *and* a poison – suggesting the undecidable lies already in the text itself, Derrida explains:

> "It has been necessary to analyze, to set to work, *within* the text of the
> history of philosophy, as well as *within* the so-called literary text [...]
> certain marks [...] that [...] I have called undecidables, that is, unities
> of simulacrum, false verbal properties (nominal or semantic) that
> can no longer be included within philosophical (binary) opposition,
> resisting and disorganizing it, without ever constituting a third term,
> *without ever* leaving room for a solution in the form of speculative

5   Derrida 1981a, 61–171.

dialectics [...]. (the *pharmakon* is neither remedy nor poison, neither good nor evil, neither the inside nor the outside, neither speech nor writing; the *supplement* is neither a plus or a minus, neither an outside nor the complement of an inside, neither accident nor essence etc.; the *hymen* is neither confusion nor distinction, neither identity nor difference, neither consummation nor virginity, neither the veil nor the unveiling, neither the inside nor the outside, etc. [...])."[6]

Derrida's doubly folded words such as *pharmakon*, *différance*, *hymen* etc., encapsulating two contradictory layers of meaning, exemplify a general *logic of the supplement*, a "neither/nor, that is, simultaneously either/or", which persists at the core of metaphysical dichotomies themselves, and erodes them from the inside out.[7]

Leaving aside, for the time being, the difficult questions concerning the status of terms such as *pharmakon*, *différance*, *supplement* or *hymen* (are they textual emblems of an overall logic of compounded difference, is the difference they set at play also textual, how do textual emblems and differences relate to a non-textual exteriority, is there indeed any so-called outside to the text at all?), I will focus instead on the fold as a case study for the differentiation implicit to the logic of the supplement. What is folded *now* operates neither as a plus nor as a minus, neither in addition to nor as a subtraction from what has already been folded *then*. It is both a plus and a minus, both an addition and a subtraction. It is neither different from what is, nor the same; rather, it is at the very same instance different and the same. It neither reveals nor conceals what has already been folded; rather, it both reveals and conceals. Imagine, for example, an ordinary sheet of paper lying on a desk. As it is being folded, the sheet of paper both increases when considering the dimensions of the embedded space and reduces when considering the space it occupies on the desk. It is still the same sheet of paper – nothing has changed with regard to the paper's chemical composition – yet it is quite different; everything has changed when considering the space it embeds and the space it is embedded in. Furthermore it not only preserves some of its main features, it also stores the potential energy of the process of transformation it underwent. It thus reveals itself as its own present and past. It also reveals other aspects, invisible up until folded, its bottom surface, for example, or flexibility, while repressing others, its former top surface, for example, or its full extension. As another example, consider the fascination invoked by the flow of drapery, the loose sagging of folds, manifest in Greek plastics and onwards in contemporary fashion, with its unresolved, ongoing play of veiling–unveiling the enveloped body. Johann Gottfried Herder in *Some observations on Shape and Form from Pygmalions Creative dream* reflects at length on the interplay between clothing and unveiling. How,

---

6   Derrida 1981b, 43.
7   Ibid.

Herder asked, could the Greek artist "clothe in such a way that nothing is hidden? Could he drape a body and yet allow it to retain its stature and its beautiful rounded fullness?"[8] Wet drapery was the answer. Only wet drapery made it possible to clothe, without veiling the body, so that drapery became, in art, what was impossible for it in actuality, a "*so to speak* drapery, a cloud, a veil, a mist."[9] Herder's reflections hinge on a distinction between the deceptive character of wet drapery, a "*so to speak* drapery, a cloud, a veil, a mist. [...] *so to speak*, just as Homers gods possess blood only so to speak," and the "fullness of the body," which in his eyes remains "the very essence of sculpture, and not merely *so to speak*."[10] With Nietzsche, however, the fullness of the body had become synonymous with the veil itself. Hence his praise for the old Greeks: "Oh, those Greeks! They knew how to live. What is required for that is to stop courageously at the surface, the fold, the skin, to adore appearance, to believe in forms, tones, words, the whole Olympus of appearance. Those Greeks were superficial – out of profundity."[11] Already in its naive manifestations – a folded sheet of paper, the sagging of folds in drapery – the fold attests to what have been and still are almost unthinkable in western philosophy: a difference beyond oppositions, unities of simulacrum rather than binaries of truth, superficiality out of profundity.

Against this backdrop, in what follows, I will attempt to delineate the extent to which thinking in folds both allows us and obliges us to re-conceptualize that which we typically signify using dichotomous dyads: one and many, subject and object, existence and essence, form and matter and so on. Within all unities of simulacra evoked by the fold – difference and identity, addition and substitution, veiling and unveiling, unity and multiplicity, inside and outside, open and closed – what interests me most, is the *double-bind* along which Deleuze develops the fold as both the impersonal machinist of the endless process of becoming, and the final cause of enclosure and finitude. On the one hand the fold will thus be described as a pure, dimensionless event that falls out of time and space, preceding every specific entity and the world in general; on the other hand the fold will figure as the curve enveloping this or that specific series of events or the world as an infinite curve in general. Is it still the same fold at both its ends? Where does the fold begin and where does it end? How to account for a world that is both given to an infinite process of becoming *and* is expressed by finite phenomena? How does the finite enter the infinite? It is in pursuit of these questions, that Deleuze, in his later writings, turned to the fold and to an affirmative reading of Leibniz, albeit in his earlier writings

---

8   Herder 1778b, 50.
9   Ibid, 51. In the german original it reads: "Das Kleid wurde in der Natur, was es in der Kunst nicht seyn kann: *gleichsam* ein Kleid, ein umhüllender Nebel." Herder 1778a, vol. 8, 137.
10   Herder 1778b, 51.
11   Nietzsche 1882, 38.

he was less sympathetic towards the so-called last polymath of European philosophy.[12] To reiterate, the double-bind develops along two complementary trajectories, a genuine differentiation, giving raise to the endless production of difference, *and* the differentiated as the finite product inevitably hiding the process of production it underwent.[13] As I will argue throughout, unfolding was Deleuze's final turn of phrase for the *pas de deux*, in which differentiation takes place while its very taking place is already covered over by the differentiated itself.[14]

III) What kind of dress for which kind of thought?

In her Neo-Baroque novel *Rachels Röckchen*, Charlotte Mutsaers develops the protagonist's portrait in a metaphorical gown vividly unfolding in numerous twists and turns. The gown's folds are described as dangling around Rachel, as spinning, wiggling, welting, blazing, flattering, crawling upon her, shimmering, crinkling, sweeping, dancing, curling, murmuring, rustling, flowing, flickering, swinging, winking, puffing up or collapsing: "the way all the folds are continuously branching out, ditching themselves or transiting from one to the other, and the way, once in a while, you catch a glimpse of what, lonely and clandestine, is happening in between or even underneath, is all that counts."[15] In *The fold* Deleuze on his part develops the texture of Leibniz's garment in its vivacious folds with their uncountable curves, swerves and inclinations. With one exception: Deleuze is not concerned with what transpires underneath the play of folds, but merely with what happens in between the folds. The fold in itself, as I will attempt to demonstrate, renders superfluous any attempt at depth beyond the surface. For what is the fold, but a paradoxical figure of transition between surface and depth?

Before discussing Deleuze's concept of the fold in detail, a few notes on his style are in order. Deleuze has explicitly pointed out the role which style plays for him within philosophy: "Becoming stranger to ones self, to ones language and nation, is not this the peculiarity of the philosopher and philosophy, or their style or what is called a

---

12 In *Spinoza et le problème de l'expression* Deleuze compares Spinoza's and Leibniz's anti-Cartisianism (Deleuze 1968). In *Difference and Repetition* Leibniz is portrayed, together with Hegel, as the philosopher of infinite representation. In Deleuze's critique of representation, Leibniz's philosophy is criticized, although a more affirmative tone concerning Leibniz's notion of vicediction is present as well. In relation to the notions of vicediction and compossibility, Leibniz once again plays an important role in another example of Deleuze's earlier work *Logique du sense* (1969). A systematic exposition of Leibniz's philosophy is ultimately presented in *The fold. Leibniz and the Baroque*. Deleuze's engagement with Leibniz can also be traced throughout his lecture series at the University of Paris in 1980 and in 1986/7. Cf. Lærke 2015, 1194–96.

13 Cf. Deleuze 1994, Chap. IV, *Ideas and Synthesis of Difference*, 168–221, especially 209–210.

14 The French expression *pas de deux* plays with the ambiguity implicit in *pas*, oscillating between negation and step.

15 Mutsaers 1997, 14 (trans. A.S.).

philosophical gobbledygook?"[16] As Deleuze and Guattari clarify in *What is Philosophy?*
style has nothing to do with rhetoric, and everything to do with "sensations: percepts
and affects, landscapes and faces, visions and becomings."[17] Style can thus be said to be
the ever-singular manner by which philosophy as the art of creating concepts encounters
its outside, the non-philosophical, the non-conceptual, the un-thought, the power that
befalls thought and forces it to think. It is the ever-singular manner by which the outside
of thought is folded into its very core and the duplicity of inside and outside finds itself
reinforced in as far as style manifests itself as thought's very own garb. It goes without
saying, neither thought nor language are ever truly naked. It goes without saying, there
is no substantial body waiting to be unveiled under their dresses and drapes. Along the
undisciplined style characteristic to Deleuze's writing *The fold* demolishes the common
practice of a well-defined line of investigation and deals with folds in all their possible
extension reaching from the folds as cosmic events to the infinite curve of the world,
from the folds in and of mathematics to the folds in and of the arts and philosophy, from
the Baroque to postmodernity, from Leibniz to Whitehead, from Caravaggio to Pollock
etc. Thus, *The fold* is certainly not to be conceived as yet another scholarly exploration
of Leibniz's philosophy, though it undoubtedly grants novel, unexpected insights into
the latter. It is first and foremost an attempt to develop a philosophical concept or an
aesthetic of the fold. And it is the fold along which the conceptual portrait of Leibniz
is drawn. Just as in a drawing or a painting, the art of the portrait is not a matter of
producing the closest possible likeness of the sitter, but the production of the resemblance itself.[18] Far from suggesting a sober, clinical reproduction, Deleuze's strategy of
a conceptual portrait implies a transformation of both sitter and artist, in this case, a
type of becoming-Deleuze of Leibniz as well as a becoming-Leibniz of Deleuze. At times
it might be hard to tell apart, which point of view is at present under consideration, is it
Leibniz's or Deleuze's? Perhaps after reading *The fold* both perspectives have become
more obscure, while at the same time the concept itself has gained in distinctness.[19] In
any event, to whom does a concept belong?

The common alignment that holds together both sides of the Leibnizian-Deleuzian becoming is the line itself, a special kind of line – a curved, or what amounts to the same, a

---

16 Deleuze/Guattari 1994, 110.

17 Ibid, 177.

18 Cf. Deleuze 1993b, 197.

19 As Deleuze argued in *Difference and Repetition*, in reference to Leibniz and his famous example of the
    murmuring sea, "distinct-obscure" or "confused and clear" are far more promising couplings than Descartes's
    "distinct and clear". "Confused and clear" and "distinct and obscure" are both called for in philosophy: the
    former as an Apollonian distinction regarding the "whole noise" of the sea and no longer being able to account
    for the little perceptions constituting it, the latter as a Dionysian distinction regarding the little perceptions and
    no longer being able to account for the "whole noise": "However, the two never unite in order to reconstitute
    a natural light. Rather, they compose two languages which are encoded in the language of philosophy and
    directed at the divergent exercise of the faculties: the disparity of style." Deleuze 1994, 213.

folded line.[20] Throughout his work Deleuze is breaking away from a certain paradigm of linearity, just as another kind of line undeniably keeps informing his thought. His writings abound in lines: "lines and speeds" is the magic formula of *A Thousand Plateaus*.[21] In his engagement with Leibniz and the Baroque, a certain point on the line, the point of inflection – the point at which a curve changes from negative downward concavity to positive upward concavity or vice versa – is crowned a cosmo-genetic element *par excellence*. Referring to Paul Klee, Deleuze identifies the point of inflection with the formers famous *Graupunkt*, a "point without dimension", a point "between dimensions", the aforementioned "locus of cosmo-genesis."[22] Thus the curved line – curves and folds are employed synonymously, to a certain extent, by Deleuze – essentially becomes active, its agent being a point in motion, every motion – an event.[23] (fig. 2)

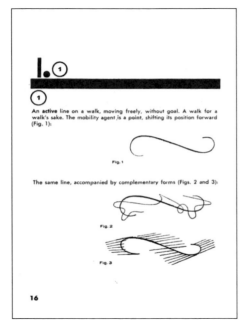

Fig. 2: Paul Klee, An active line on a walk.

20 Though Deleuze's reference to passages, in which Leibniz explicitly uses the terms 'fold' or 'folding', are rather scarce, the fold nevertheless and without a doubt constitutes the main interpretive move both for his reading of Leibniz and his understanding of the Baroque. His most important references are to Leibniz's *Pacidius to Philaletes* (Leibniz 1676, 145); *Protogaea* (written between 1791–1793), see Leibniz 2008, chap. VIII, 20–25; *Die philosophischen Schriften*, see Leibniz 1965a, IV, 481–482; VI, § 61, 617; VII, 453. Deleuze's intuition, regarding the role the fold plays in Leibniz's texts, has been affirmed – even though Deleuze could not have foreseen that – by later publications out of Leibniz's oeuvre. Cf. for ex. Leibniz 1999, VI, 1401, 1687, 1900. Cf. Lærke 2015, 1197–98.
21 Deleuze/Guattari 1987, 4.
22 Deleuze 1993a, 30. Cf. Klee 1956, 3–4.
23 Cf. Klee 1953, 16. By no means accidental, Klee also portrays the active line as an S-shape in reverse, reminiscent of the *figura serpentinata*, which, particularly in mannerism, played a decisive role with respect to the disturbance or distortion of classical forms. Cf. Uhlig 2007, 307.

As the beginning of the world the event of the fold in a way falls out of time and space and as such only gives rise to the dimensions and coordinates constituting the quantitative space-time, in which we are moving, practically and theoretically, from one place, from one moment, from one topic to another. It is against the backdrop of the event that Deleuze breaks with the paradigm of linearity in order to replace one kind of line with another – the straight line of the classical age with the curved line or fold of the Baroque – a substitution which also exchanges one kind of philosopher for another, and thus exchanges two types of reason: René Descartes's with Leibniz's. Concerning the twofold labyrinth within which both Descartes and Leibniz find themselves wandering, "the continuous labyrinth in matter and its parts, the labyrinth of freedom in the soul and its predicates," Deleuze writes:

> "If Descartes did not know how to get through the labyrinth, it was because he sought its secret of continuity in rectilinear tracks, and the secret of liberty in a rectitude of the soul. He knew the inclension of the soul as little as he did the curvature of matter. A cryptographer is needed, someone who can at once account for nature and decipher the soul, who can peer into the crannies of matter and read into the folds of the soul." [24]

While all straight lines resemble each other, the curved line or fold implies infinite variation. Every fold takes on a different course, just as no two things – leaves, rocks, rivers, drops of water, etc. – are folded in the same way, not one regular fold pervades one and the same thing. With Leibniz and Deleuze the fold is everywhere and nowhere the same. Hence, the fold must not be perceived as universality, but rather as a universal differentiator. From Deleuze's perspective, it is taking the divergent path, preferring the swerve to the straight line that lends Leibniz's conceptual portrait its specific baroque traits. With "inclension of the soul" and "curvature of matter" Deleuze points at the main characteristics, which in his eyes allow for an approximation to the Baroque in and beyond Leibniz. [25] Unsurprisingly, Deleuze is not interested in contributing to the debates regarding the history of style or the epochal concept of the Baroque. His interests are rather directed at an elaboration of what he identifies as the "operative function" of the Baroque, a function that consists in endlessly producing folds, pushing the folds to infinity "fold over fold, one upon the other." [26]

---

24 Deleuze 1993a, 3.
25 Ibid.
26 Ibid.

IV) Falling drapes and folded tableaus

To be sure, the fold is not a genuine invention of the Baroque. Much of art history could be portrayed – and Georges Didi-Huberman did indeed attempt to do so in *Ninfa moderna* – as the history of folds or falling drapes.[27] Here too, it all begins with something going quite radically off-course; *clinamen* is the word Didi-Huberman borrows from the Latin philosopher and poet Lucretius (1st c. B.C.E.) to express the pervasive obliqueness of things.[28] In his philosophical and didactic poem *De rerum Natura* Lucretius employs the term *clinamen* to translate Epicurus' *parenklisis*, the indiscernible motion by which atoms are thought to be veering minimally from free fall. Whilst hurtling straight down through empty space, the atoms simultaneously diverge from their path, through some impetus of their own; moreover, they do so at an angle and speed that can neither be comprehended nor imagined.

> "The atoms, as their own weight bears them down / Plumb through the void, at scarce determined times, / In scarce determined places, from their course / Decline a little – call it, so to speak, / Mere changed trend. For were it not their wont / Thuswise to swerve, down would they fall, each one, / Like drops of rain, through the unbottomed void; / And then collisions ne'er could be nor blows / Among the primal elements; and thus / Nature would never have created aught."[29]

Because atoms decline from parallel paths, they hit each other. Collisions and blows, cosmic turbulences, are the result, ultimately leading to the metastable systems of different worlds. Didi-Huberman refers to the Lucretian *clinamen* in order to track the long history of the falling drape in a kind of cinematographic documentary, consolidating the innumerable swerves of the fold in European art history. Through the eyes of Aby Warburg, through a modern science of the image, Didi-Huberman lets his movie of falling drapes depart from the motif of the nymph. Nymphs: "wonderfully draped apparitions which come from who knows where; prancing in the wind, always touching, not always well-behaved, almost always erotic, sometimes disturbing."[30] Using the examples of the

---

27 Didi-Huberman 2006.

28 Lucretius' *De rerum Natura* (DRN) figures not only as a translation of Epicurus's philosophy of nature, but also as one of the most important and elaborated documents of materialism and atomism of antiquity. In the DRN the world as such is thought of as an infinite material texture (*textura rerum*), the atoms being the elements out of which the texture is woven by their own spontaneity (*sponte sua*). While Epicurus defined them negatively as indivisible bodies, ἄτομοι, Lucretius characterizes them additionally as seeds of things (*semina rerum*) and generative bodies (*genitalia corpora*) to underline their formative aspects. He also calls them blind / invisible bodies (*corpora caeca*) in order to emphasize the aimless and chaotic dynamic they are involved in and set at play. The atoms' aimless and chaotic nature refers back to the above-introduced *clinamen*. Cf. Moser 2014, 5.

29 Lucretius, De rerum natura (1st c. B.C.E.), in: Lucretius 1994–2000, II, v. 217–224.

30 Didi-Huberman 2006, 11.

so called *florentine nymph* in Ghirlandaio's *Birth of Johannes*, as well as in Botticelli's *Allegory of Spring* and *Birth of Venus*, Warburg exposed the *afterlife of antiquity* throughout renaissance and humanism, paying special attention to the displacement of the *pathos formula* from the figure to its edges, to hair and the folds of cloth fluttering in the wind, to the mobile accessories (*bewegtes Beiwerk*).[31] (Fig. 3)

Fig. 3: Domenico Ghirlandaio, "The birth of Johannes", 1486–1490, fresco, S. Maria Novella, Cappella Tornabuoni, detail.

On the verge of modern representation, Didi-Huberman asserts that Warburg's nymph would not only have "slowed her pace" but would finally have tumbled over. (Fig. 4, fig. 5) Between the decline of drapes and the fall of the nymph Didi-Huberman marks an alignment and resonance that is expressed in the slow detachment of body and cloth, of nudity and that which envelopes it – a subtraction with remainder. What remains, Didi-Huberman argues, is drapery itself, a piece of clothing that has slid to the ground, a rag, arriving finally at the runnels of the modern European city. From Moholy-Nagy's *Trottoirs* up to the folds in felt by Robert Morris, (fig. 6, fig. 7) they all embody – and here Didi-Huberman refers to Deleuze – the possibility inherent to art of positing form as folded. In a nutshell, the *clinamen* implicit to matter would already have instigated the subversion and decline of ideal forms, a decline that does not lead to the negation of form, but to another conception of form, namely, to form as folded.[32]

---

31 Cf. Warburg 1893.
32 Deleuze 1993a, 35; Cf. Didi-Huberman 2006, 135.

Fig. 4: Tizian, "Bacchanal", 1518/19, oil on canvas, 175 cm x 193 cm, Madrid, Museo del Prado.

Fig. 5: Nicolas Poussin, "The triumph of Pan", 1636, oil on canvas, 138 cm x 157 cm, London, National Gallery.

Fig. 6: Lászlo Moholy-Nagy, "Rinnstein", 1925, gelatin silver print, 28,9 cm x 20,8 cm.

Fig. 7: Robert Morris, "Untitled (Emmêlement, Tangle)", 1967, felt, New York, Museum of Modern Art.

With Didi-Huberman we took a glimpse at the folds within painting, at the double decline of the nymph and drapery, a fall that ends, as we have seen, in the runnels of Modernity.[33] With Deleuze we once again rewind Didi-Huberman's movie on folds and veer a little from the path undertaken: we swerve from the fold in painting to the painting as a folded tableau. With the folded tableau *(tableau ployant)* I refer to a notion introduced by Hubert Damisch to express the formal – folded – structure of the *Baroque Narcisse*.[34] (fig. 8) The painting in question is characterized by a horizontal fold, which divides the plane of the picture into a lower and an upper half. Both as connecting and as separating the two halves, the operation of the fold thus encompasses the folded totality of the plane as such. Years before Damisch's notion of the folded tableau found its way into art history discourse, Deleuze, similarly though in less detail, described the operation of an internal pictorial fold characteristic of otherwise incomparable Baroque painters such as El Greco and Tintoretto. Looking at El Greco's *The Burial of Count Orgaz* (fig. 9) Deleuze focuses on the horizontal line splitting/duplicating the painting into a lower and an upper half; in the lower half "bodies are pressed leaning against each other," while in the upper half "a soul rises along a thin fold, attended by saintly monads, each with its own spontaneity."[35]

---

33 Didi-Huberman 2006, 31.

34 Damisch 1996, 33.

35 Deleuze 1993a, 30.

Fig. 8: Michelangelo Merisi Caravaggio, "Narcissus", 1608–1610, oil on canvas, 113 cm x 97 cm, Rome, Palazzo Corsini, Galleria Nazionale d'Arte Antica.

Fig. 9: El Greco, "The burial of Count Orgaz", oil on canvas, 1586–1588, 480 cm x 360 cm, Toledo, Santo Tomé.

In Tintoretto's *Last Judgment* Deleuze detects the same operation of splitting/duplicating the totality of the pictorial plane into a lower and an upper half and as such constituting its totality (fig. 10):

> "In Tintoretto the lower level shows bodies tormented by their own weight, their souls stumbling, bending and falling into the meanders of matter; the upper half acts like a powerful magnet that attracts them, makes them ride astride the yellow folds of light, folds of fire bringing their bodies alive, dizzying them, but with a dizziness from on high (un vertige du haut): thus are the two halves of the Last Judgment."[36]

In alignment with Heinrich Wölfflin, Deleuze characterized the world of the Baroque as extended across two axes – "a deepening toward the bottom, and a thrust toward the upper regions" – the first physical, concerning bodies in their materiality, the second metaphysical, concerning souls and their freedom.[37] Both are separated and, at the same time, held together by one and the same operation: the operation of folding.

Fig. 10: Jacopo Tintoretto (copy after), "Last judgement", 17th century, oil on canvas, Venice, Museo Correr.

---

36   Ibid.
37   Ibid, 29.

V) The Baroque house of thought

In the context of Deleuzian thought the fold is considered, first and foremost, in its opera-tive function. Attention is granted to the manner by which certain relations are articulated, the relations between: top and bottom, inside and outside, material and immaterial and others. The meaning of the fold transcends its phenomenological manifestations and reaches into the realms of mathematics, physics, epistemology and metaphysics. In Deleuze's reading, the fold affords a path towards a different concept of thinking and an equally altered concept of the world as such; a world affected, compressed and curved by the interplay of forces and matter.[38] As previously suggested, the fold is the thread along which Deleuze proceeds through the labyrinth of the Leibnizian legacy. A legacy that folds or rather doubles over splitting itself into two infamous labyrinths, as Leibniz says: "one is the great question *of freedom and necessity* [...]; the other is the debate on *continuity* and *indivisible things* [...]."[39] Drawing on Leibniz bipartite differentiation, Deleuze's allegory of the Baroque house of thought likewise presents itself as a diptych. It depicts the division of one house into two floors, a lower and an upper one. The labyrinthine continuum of matter and its constituents is located on the lower floor, while the labyrinth of the soul is located on the upper floor. Contrary to the Platonic distinction between two worlds, in contrast also to the model of ascension in the neoplatonic tradition, the Baroque house of thought knows only one world with two floors, separated and held together by a single fold "that echoes itself, arching from the two sides according to a different order. It expresses [...] the transformation of the cosmos into a mundus."[40]

In and between the two floors everything is happening according to the operations of the fold. The connection/separation between the two in itself is produced through folding. The fold, as both the inner folds of the soul (*plis*) and the outer pleats of matter (*replis*), marks their difference *and* their oneness, marks their differential affiliation with one and the same world. The world depicted as the common house of matter and soul, is thus made up of two infinite series of folds, one series unfolding – realizing – the pleats of matter, the other one unfolding – actualizing – the folds of the soul. I will later touch on just how realization belongs to matter and actualization belongs to the soul, while the subject of actualization and realization is the virtual. Within Deleuze's thought, the virtual is the domain of the ideal or problematic. Already in *Difference and Repetition*, Deleuze adopts the notion of the fold as expressing the relation between the virtual with its actualization and realization in terms of implication, explication and complication. The virtual – in the following quote designated as chaos – implicates the genetic elements, which will eventually be by explicated, i.e., actualized and realized. Deleuze writes:

---

38   Ibid, 45
39   Leibniz 1985, § 189.
40   Deleuze 1993a, 29.

"The trinity complication-explication-implication accounts for the totality of the system – in other words, the chaos which contains all, the divergent series which lead out and back in, and the differenciation [the fold] which relates them one to another. Each series explicates or develops itself, but in its difference from the other series, which it implicates and which implicates it, which it envelops and which envelops it; in this chaos which complicates everything. The totality of the system, the unity of the divergent series as such, corresponds to the objectivity of a problem." [41]

Within the trinity of complication-explication-implication the infinite curve of the world is set at play: "*The world is the infinite curve that touches at an infinity of points an infinity of curves* [...]". [42] And the whole world – thus the important addition – is "enclosed in the soul from one point of view". [43] Nevertheless, and this should become clear in what follows, the virtual, its actualization and realization are not to be understood as reciprocally exclusive, but as strictly complementary. Before entering the discussion on their mutual unfolding, let us first take a closer look at Leibniz's concepts of the labyrinth of the continuum of matter and the labyrinth of freedom in the souls, each on its own terms and in relation to the fold.

(i) The external folds *(replis)* of matter

It is specifically in reference to the labyrinth of the continuum of matter that Leibniz refers most explicitly to the fold, as in *Pacidius to Philaletes*, a text dating back to 1676 that deals with the problem of the continuum in its physical sense. Initially, the atomists' radical solution, as well as Descartes's definition of matter in terms of extension, is left aside. [44] In contrast to the microscopic discontinuity that underlies the sense-based experience of continuous matter as postulated by the atomists, Leibniz allows only for gradual differences, both in relation to the divisibility of material bodies and their motion.

---

41 Deleuze 1994, 123–4.
42 Deleuze 1993a, 24.
43 Ibid.
44 Despite the broad spectrum of divergent points of view characterizing early modern atomism, itself referring back to the renaissance of the notion of atomism in antiquity, and especially to the rediscovery of Lucretius, a general affirmation of the composition of complex bodies out of naturally indivisible material atoms, can be asserted. While Leibniz sympathized with such a conception of the body in his early physics, in his later physics he insisted on its incompatibility with his general understanding of the natural world. A similar development can be diagnosed in regard to Leibniz's account of the mechanical philosophy of his time. While a certain sympathetic affiliation with the new mechanical philosophy and its aim to explain all natural phenomena in terms of matter and motion prevails throughout his work, his later physics should at the same time be read as a thorough critique of the mechanistic tradition. Cf. McDonough 2014, ch. 2.

Instead of a cluster of primary indivisible particles, the fold as a relevant concept enters the discussion; instead of a perfectly solid or perfectly fluid body, there appears an elastic yet resistant body.

> "I myself admit neither Gassendi's atoms, i.e. a body that is perfectly solid, nor Descartes subtle matter, i.e. a body that is perfectly fluid [...] the division of the continuum must not be considered to be like the division of sand into grains, but like that of a sheet of paper or tunic into folds [...]. It is just as if we suppose a tunic to be scored with folds multiplied to infinity in such a way that there is no fold so small that it is not subdivided by a new fold: and yet in this way no point in the tunic will be assignable without it being moved in different directions by its neighbors, although it will not be torn apart by them. And the tunic cannot be said to be resolved all the way down into points; instead, although some folds are smaller than other to infinity, bodies are always extended and points never become parts, but always remain mere extrema."[45]

Neither perfectly solid, nor perfectly fluid, Leibniz conceives of matter as an elastic, continuous and endlessly folded texture. Folded into ever-smaller folds, matter does not ever break down into primary atomic constituents, nor is its cohesion (tension and release) ever lost. Matter thus constitutes an infinite continuum, wherein "no point [...] will be assignable without it being moved in different directions by its neighbors, although it will not be torn apart by them."[46] Contrary to the atomists' hypothesis, matter is thus perceived as indecomposable into primary particles. Matter rather forms a variety of masses according to the motion and the forces active in and between its folds. This takes us into Leibniz's complicated ontology of forces. Although this is not the place to give a detailed account of it, a few basic remarks are inevitable to shed some light on the terms which Deleuze uses to identify two basic types of forces: an *elastic-compressing force* responsible for the accumulation of matter along its outer, inorganic pleats, and a *plastic force*, responsible for the organization of matter along its inner, organic pleats.[47]

Leibniz's critique of atomism led, as we have seen, to a conception of material bodies as irreducibly elastic and infinitely divided into ever-smaller folds. The material world is thus conceived as a worldwide net of ever-smaller folds. In and between the worldwide

---

45   In: Leibniz 2001, 185, in the English translation, the word "extrema" is used, while Leibniz uses "Grenzen". Cf. Leibniz 1676, 145.

46   Leibniz 2001, 185.

47   This diagrammatical sketch is developed in the 1st chapter of *The fold*. See Deleuze 1993a, 3–13.

net of material folds, different forces are at play. Leibniz divides them into passive and active, into primitive and derivative. He writes:

> "*Active force* is twofold, that is, either *primitive*, which is inherent in every corporeal substance *per se* [...] or *derivative*, which, resulting from a limitation of primitive force through the collision of bodies with one another, for example, is found in different degrees. Indeed, primitive force (which is nothing but the first entelechy) corresponds to the *soul* or *substantial form*. [...] Similarly, passive force is also twofold, either primitive or derivative. And indeed, the *primitive* force *of being acted upon* [*vis primitiva patiendi*] or of resisting constitutes that which is called *primary matter* in the schools, if correctly interpreted. This force is that by virtue of which it happens that a body cannot be penetrated by another body, but presents an obstacle to it, and at the same time is endowed with a certain laziness, so to speak, that is, an opposition to motion, nor, further, does it allow itself to be put into motion without somewhat diminishing the force of the body acting on it. As a result, the *derivative force of being acted upon* later shows itself to different degrees in *secondary matter*."[48]

In what follows the passage quoted above, the two facets of an active force are spelled out as an elementary *dead force* (*vis mortua*) restricted to the initiation of motion and an ordinary *live force* (*vis viva*) joined with actual motion.[49] As examples of dead forces Leibniz lists: the centrifugal force, the force of heaviness and the force that restores a stretched elastic body back to its original state. The living force is assigned to the impact arising from the fall of heavy bodies. Passive forces, contrary to active force, are related to the resistance to motion: "a force [...], an inclination to retain its [a thing's] state, and so to resist changing."[50] In what concerns the difference between primitive and derivative forces, the primitive active force is assigned to the soul or substantial form, while the primitive passive force is assigned to primary matter. Together they complete a corporeal substance. The derivative forces, on the other hand, are those commonly investigated by physicists analyzing size, shape and motion of natural phenomena in as far as they satisfy certain laws.[51] In Leibniz's opinion, only with regard to primitive forces – that is with regard to the primary activity of the soul or substantial form – an escape from the dead ends of mechanistic reductionism is imminent. Leibniz's critique of the mechanistic

---

48 Leibniz, *Specimen Dynamicum* (1695). In: Leibniz 1849–63, vol. VI, 236–7; Leibniz 1989, 119–20.
49 Ibid, 238; 290.
50 Leibniz, *Letter to de Volder* (March 24th/April 3rd, 1699). In: Leibniz 1965a, vol. II, 170; Leibniz 1989, 172.
51 Leibniz, *Specimen Dynamicum* (1695). In: Leibniz 1849–63, vol. VI, 237; Leibniz 1989, 120.

approach does not lead to its negation, but to a reconsideration of its limits. As Leibniz puts it in *Discourse on Metaphysics*:

> "Although all particular phenomena of nature can be explained mathematically or mechanically by those who understand them, nevertheless the general principles of corporeal nature and of mechanics itself are more metaphysical than geometrical, and belong to some indivisible forms or natures as the causes of appearances, rather than to corporeal mass or extension."[52]

As long as we move from fold to fold along the creases introduced by derivative forces and explained by mathematics and mechanics according to Leibniz we will ultimately be restricted to perceive the unfolding of the world only in mathematical and mechanical terms and thus miss its real nature. Let my try to make this point as clear as possible: Leibniz holds on to the hypothesis that matter forms an infinite continuum and consequently denies the existence of any final indivisible point, which would allow us to determine the limits of a specific body or motion. Then, the question arises: in what way should a discernable unity within the infinite multiplicity of matter and motion be conceived. For Leibniz the specific unities of matter and motion point, as we have seen, to "some indivisible forms or natures", transcending the mathematical and mechanical realm.[53]

With this in mind we can return to Deleuze and his differentiation between elastic-compressing forces as assigned to the accumulation of matter along its outer, inorganic pleats and his so called plastic forces, assigned to the organization of matter along its inner, organic pleats. Both the elastic-compressing forces and the plastic forces must be conceived as derivative forces in Leibniz's sense. In Deleuze's reading of Leibniz one and the same worldwide net of material folds is thus developed along two complementary lines: under the influence of elastic-compressing forces, matter, as a mass, forms external folds that encompass an *outer milieu*. The outer milieu may be considered inorganic, to an extent. Subject to plastic forces, matter, as an organism, forms inner folds that encompass an *inner milieu*. The inner milieu may be considered organic, to an extent. 'To an extent' means that there is no difference in essence between the inner and the outer, between that which is organic and that which is inorganic, but only "a difference

---

52 Leibniz, *Discourse* (1686). In: Leibniz 1965a, vol. IV, 444: Leibniz 1989, 51–2.
53 Deleuze certainly follows Leibniz's anti-mechanical trend, by pushing it towards a different end: a mechanism for Deleuze is no longer faulty "for being too artificial to account for living matter, but for not being mechanical enough, for not being adequately machined." Deleuze 1993a, 8.

of vector".[54] Leibniz himself uses the images of a garden full of plants or a pond full of fish to express the intertwined logic of inside and outside, organic and inorganic: "Every portion of matter may be conceived as a garden full of plants, and as a pond full of fish. But every branch of each plant, every member of each animal, and every drop of their liquid parts is in itself likewise a similar garden or pond."[55] An organic inside is thus inhabiting an inorganic outside, as in the image of fish in a pond, while each fish in itself is once again a host to a throng of living being and thus functions as the outside to the latter. "The inorganic folds," as Deleuze puts it, "move between two organic folds. For Leibniz, as for the Baroque, the principles of reason are veritable cries: Not everything is fish, but fish are teeming everywhere."[56]

Matter, located on the bottom floor of the Baroque house of thought, thus, turned out to be folded into masses and organisms, into accumulations and living beings, into outer and inner milieus, according to the interplay of elastic-compressing and plastic forces. Recall that both the elastic-compressive and the plastic forces were considered derivative forces and as such insufficient to account for the unity of corporeal bodies and motion. For Leibniz, the unity of corporeal bodies and the unity of motion, necessarily point to another "higher inner and individualizing entity,"[57] to the reign of primitive forces and hence to the labyrinth of the soul.

(ii) The internal folds (plis) of the soul

Changing floor, traversing from the labyrinth of matter to the labyrinth of the soul, implies a shift from Leibniz's physics towards his metaphysics. With the introduction of primitive forces, this shift is already noted. Just as his physics, Leibniz's metaphysics too can be divided into an early and a late period. The early period is generally regarded as reaching from Leibniz's youth to the Discourse on Metaphysics (1686), while the late period is generally considered to expand from the New System (1695) to the theory of monads developed since then.[58] The theory of monads constitutes the main focus of the following remarks. Since its very first articulation, Leibniz's theory of monads has provoked a wide range of interpretations, trying to come to terms with the unseizable description of the monad as a

---

54 Ibid, 8. In its genealogy the concept of milieu can be traced back to 18th century biology. From then on it played a fundamental role in the historically diverging conceptions of the living individual in relation to its environment. In A thousand Plateaus Deleuze and Guattari mainly refer to Jakob v. Uexküll and Gilbert Simondon to further elaborate on a contemporary concept of the milieu. See Deleuze/Guattari 1987, especially the chap. titled The Geology of Morals, 39–75.

55 Leibniz 1965b, § 67.

56 Deleuze 1993a, 9.

57 Ibid.

58 For an introduction to Leibniz's early and late metaphysics see Mercer/Sleigh 1994; Rutherford 1994.

simple, soul-like substance, unextended, without parts and without windows,[59] and more generally, with the integration of the theory of monads within Leibniz's account of matter and the interplay of body and soul.[60] In *The fold* Deleuze provides a seminal reading of Leibniz's theory of the monads. Within the limits of this paper neither Leibniz's original theory nor Deleuze's interpretation can be developed to their full extent. I would rather restrict myself to stressing the foundational role that Leibniz assigns to the substance as a principle of force and of true unity and to sketching its relation to the concept of the fold. To be sure, Leibniz's notion of the monad as unextended, soul-like substance implies a radical reconceptualization of the notion of substance as such. As we have seen above, Leibniz saw it necessary to reintroduce the notion of substantial forms in order to account for the unity of corporeal bodies and of motion. The reintroduction of substantial forms does not come without a transformation of substance in terms of forces. Leibniz writes: "[...] it was necessary to restore, and, as it were, to rehabilitate the *substantial forms* [...], but in a way that would render them intelligible, and separate the use one should make of them from the abuse that has been made of them. I found then that their nature consists in force, [...]."[61] It is due to this de-substantialization of the substance in terms of forces that Deleuze characterizes Leibniz (and the Baroque in general) as exchanging the hyle-morphic model for the material-force model.[62]

Following Deleuze, I will from now on traverse a path, which according to him connects the three essential phases defining the internal folds of the soul, from *inflection* to the *point of position* (or *point of view*) and from the latter to the *envelope of inherence* (or *inhesion*).[63] (fig. 11)

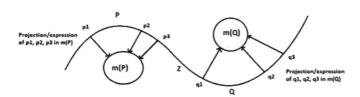

Fig. 11: Point of inflection (Z), point of view (P, Q), point of inclusion (m(P), m(Q)).

59   Cf. Leibniz 1965b, § 1 and § 7.
60   Cf. Rutherford 1994, 124. Bredekamp 2008 offers an exceptional reading of Leibniz and the windows of the monad.
61   Leibniz, *New System* (1695). In: Leibniz 1965a, vol. IV, 478–79; Leibniz 1989, 139.
62   Cf. Deleuze 1993a, 35.
63   Cf. Deleuze 1993a, 20–22. The corresponding French and German terms are: *point d'inflexion/Inflexionspunkt, point de vue/Gesichtspunkt* and *enveloppe d'inhérence, d'inhésion/Hülle der Inhärenz, Inhäsion.* Cf. Deleuze 1988a, 28–31.

Before doing so, let me briefly address a more general question: what do the *internal folds* of the soul stand for? Recall the partition into an upper and a bottom floor that characterizes the Baroque house of thought. On the bottom floor we posited the labyrinth of matter, on the upper floor the labyrinth of the soul. In reference to Leibniz and his description of the monad as simple substance without windows, Deleuze depicts the upper floor as some kind of "dark room or chamber decorated only with a stretched canvas 'diversified by folds,' as if it were a living dermis."[64] Though unextended, without parts and without windows, the unity of substance is in itself diversified. Leibniz describes the internal diversification as "a multiplicity within the unity of the simple substance", as a "plurality of states and of relationships", consisting in nothing else but "what we call perception".[65] "And that is all that can be found in a simple substance –", Leibniz continues, "perceptions and changes of perceptions [appetition]."[66] According to Leibniz the perceptions placed on the opaque canvas of the soul, represent an innate form of knowledge; Deleuze on his part describes them in terms of ideal events. Now, what is an ideal event? According to Deleuze an ideal event is a set of singularities, mathematically speaking, a set of singular points that characterize a curve.[67] In the terminology of *The fold* the ideal event corresponds to the point of inflection. A second distinguishing point between Leibniz and Deleuze should be emphasized: while according to Leibniz a pre-established harmony precedes individual substances, for Deleuze no such harmony could be supposed. For Deleuze a pre-individual field is thought to precede the individual, namely, the virtual. It is once more in analogy to mathematics, more precisely, in analogy to the mathematical concept of a manifold that Deleuze specifies the notion of the virtual as a pure *Many*, as a purely disjunctive diversity.[68] In order for the ideal event to pass from a state of mere virtuality into actuality and reality, it must be actualized by the soul and realized by matter. Having made this point, we can now return to the figure introduced above. The point of inflection (Z) designates the formative force, the ideal event of the fold – the singularities of perception. The points of view (P), (Q) designate the vectors of curvature (p), (q), which indicate the direction of concavity and stand for a place, a site or a position – the appetite that leads the perceptions to change. Lastly, the points of inclusion m(P), m(Q) designate what insists in the point of view: the soul or the subject as "an envelope of inherence or of unilateral 'inhesion'".[69] Inclusion, inherence or inhesion, Deleuze argues, "is the final cause of the fold".[70] Deleuze's line of thought thus becomes apparent: instead of starting off with the soul or subject, he starts from the event of the fold, as the formative force giving rise to a series of subsequent transfor-

---

64   Deleuze 1993a, 4. Cf. Leibniz 1996, II, chap. 12, 114ff.

65   Leibniz 1965b, § 13 and § 14.

66   Ibid. § 18.

67   Cf. Deleuze 1969, 67.

68   Deleuze 1993a, 76.

69   Ibid, 41.

70   Ibid.

mations. Infinitely many points of views are generated by infinitely many events of folding, transforming concavity to convexity and vice versa, points of view, which than figure as points of position for a soul or subject to envelop. Nevertheless, the process which I have delineated cannot be conceived as one-directional, for "why would something be folded, if it were not to be enveloped, wrapped, or put into something else?"[71]

Let me linger a while longer on inflection itself. In Deleuze's terms inflection was defined as the ideal event happening to the infinite curve of the world, such that the *"infinite curve that touches at an infinity of points an infinity of curves"*[72] must be considered the ever-unfinished product of the events of inflection. At the heart of these considerations is the notion of a thoroughly eventful world, within which every inflection becomes the locus of a new fold, pushing the variation towards infinity: "That is how we go from fold to fold and not from point to point, and how every contour is blurred to give definition to the formal powers of the raw material, which rise to the surface and are put forward as so many detours and supplementary folds."[73] According to Deleuze every inflection is to be considered a variation pulverizing the entire world into an infinite number of ever-smaller folds. But the pulverization of the world does not come without the folds simultaneously exceeding themselves or finding their finality in an inclusion.[74] What formerly seemed an end result to the process of unfolding – the soul or subject – is, then, at the same instance designated the final cause of folding. Nevertheless, between the process of unfolding and the inherence of the soul or subject an essential gap or asymmetry is present, an asymmetry between the virtuality of the event of folding (point of inflection) and the actuality of inherence (point of inclusion). "What is folded", Deleuze concludes, "is the included, the inherent," furthermore "it can be stated that what is folded is only virtual and currently exists only in an envelope, in something that envelops it."[75] By enveloping the infinite curve of the world, the soul or subject actualizes it. But something is still missing to complete the picture we have been drawing: realization as belonging to matter. In a nutshell, the Leibnizian-Deleuzian allegory of the Baroque house of thought is to be considered a single virtual, infinitely curved plane, actualized by the souls or subjects on the upper floor and realized by matter on the bottom floor. Rather than maintaining a horizontal divide, a suitable image for the Leibnizian-Deleuzian allegory would be one that folds one floor on top of the other into a single plane. Every fold of this plane would constitute a thin membrane alongside which the infinite outside of the continuum of matter and the essential enclosure or finitude of every actual being touch upon each other, actualizing and realizing the infinite potential of the virtual plane.

---

71   Ibid, 22.
72   Ibid, 24.
73   Ibid, 17.
74   Cf. ibid.
75   Ibid.

VI) The inside of the outside

What does thinking in folds ultimately imply? It implies a philosophy of the event and a corresponding theory of differentiation and individuation. The differentiated and the individuated are no longer conceived as miraculously presupposed or as deduced from some kind of ideal form, but as the actualization and realization of a common virtual plane. Whatever appears in front of our eyes necessarily presents itself to us as that which has already been actualized and realized, as that which has already been differentiated and individuated, as that which has already been unfolded. To a degree, the process of differentiation and individuation – the process of unfolding – will always be buried beneath some assumed form; that is to say, the process itself is always in danger of being covered by its own products and thus of being overlooked.[76] The differentiated will inevitably present itself to us in a certain form and as having certain qualities, which necessarily veil its initially formless and unqualified intensities: "In brief, we know intensity only in the extended fold and the object veiled in qualities."[77] By analogy with Paul Klee, for whom the creative forces themselves cannot be named, differentiation can be said to remain unnamable to an extent.[78] Nevertheless, as Klee continues, the creative forces do reveal themselves and have to be revealed in the known types of matter, just as the curved line reveals itself as a trace of the virtual intensities preceding its actual form. In reference to Henri-Louis Bergson, Deleuze considers the virtual plane to involve the entire past – not only the past conserved in the actual form, and the future that will at some point be actualized, but also a past that was never present, and a future that will never become present.[79]

Following in Derrida's footsteps we have initially posited the fold as another exemplary figure of the logic of the supplement expressed in the syntactical form of "neither/nor, that is, simultaneously either/or".[80] Like Derrida's doubly-folded words, which subvert the binary order characteristic of western metaphysics from within the metaphysical text itself, the fold has been shown to unsettle the dichotomies between now and then, veiling and unveiling, difference and identity, organic and inorganic, matter and soul, virtuality, actuality and reality. The intimate duplicity of folding – "a severing, by which each term casts the other forwards, a tension by which each fold is pulled into the other"[81] – unsettles the very foundation of ontology as such, namely the spatial, temporal and normative order of being as presence. For both Deleuze and Derrida – choosing different points

---

76 Cf. Deleuze 1994, 281.

77 Ibid, 282–283.

78 In *Die Kraft des Schöpferischen* Klee states: "Die Kraft des Schöpferischen bleibt letzten Endes geheimnisvoll. [The creative forces ultimately remain inexplicable.]", Klee 1956, 17.

79 Cf. Deleuze 1994, 82ff.

80 Derrida 1981b, 43.

81 Deleuze 1993a, 30.

of view, using different conceptual tools and arriving at different conclusions – thinking in folds amounts to an ontological revolution or rather to a substitution of the logic of being for a logic of difference. With his notion of differentiation Deleuze substitutes the Platonic concept of a pure being-without-becoming for a pure becoming-without-being. With the notion of différance Derrida substitutes the ontological question "what is?" for an affirmation of the trace of a radical alterity, subverting both ontology's interrogative form and the authority of presence and identity underlying it. Deleuze's differentiation, as much as Derrida's différance, are to be conceived the other to "all that is". But the other can no longer be located simply beyond or outside "what is" – for if it were the case, a mere inversion would once again put the order of being upside down, while still maintaining the same ontological frame. In order to poke a hole into the frame itself, in order to open it up for "all that is not", the other or outside must be conceived as insisting within all the folds and foldings which together make up this or that inside: "they are not something other than the outside, but precisely the inside *of* the outside."[82]

---

82   Deleuze 1988b, 97.

Bibliography

Bredekamp, Horst (2008): *Die Fenster der Monade. Gottfried Wilhelm Leibniz' Theater der Natur und Kunst*. Berlin: Akademie Verlag.

Damisch, Hubert (1996): *Narcisse Baroque?*. In: Buci-Glucksmann, Christine (ed.): Puissance du Baroque. Paris: Éditions Galilée, pp. 29–43.

Deleuze, Gilles (1968): *Spinoza et le problème de l'expression*. Paris: Minuit.

Deleuze, Gilles (1969): *Logique du sens*. Paris: Minuit.

Deleuze, Gilles (1988a): *Le pli. Leibniz et le Baroque*. Paris: Minuit.

Deleuze, Gilles (1988b): *Foucault*. Trans. and ed. by Hand, Seán. Minneapolis: University of Minnesota Press.

Deleuze, Gilles (1993a): *The Fold. Leibniz and the Baroque*. Trans. by Conley, Tom. London: The Athlone Press.

Deleuze, Gilles (1993b): *Unterhandlungen: 1972–1990*. Trans. by Roßler, Gustav, Frankfurt a.M.: Suhrkamp.

Deleuze, Gilles (1994): *Difference and Repetition*. Trans. by Patton, Paul. New York: Columbia University Press.

Deleuze, Gilles / Guattari, Félix (1987): *A Thousand Plateaus. Capitalism and Schizophrenia*. Trans. by Massumi, Brian. Minneapolis: University of Minnesota Press.

Deleuze, Gilles / Guattari, Félix (1994): *What is Philosophy?*. Trans. by Burchell, Graham / Tomlinson, Hugh. London / New York: Verso.

Derrida, Jacques (1981a): *Dissemination*. Trans. by Johnson, Barbara. Chicago / London: The University of Chicago Press.

Derrida, Jacques (1981b): *Positions*. Trans. by Bass, Alan. Chicago / London: The University of Chicago Press.

Didi-Huberman, Georges (2006): *Ninfa moderna. Der Fall des Faltenwurfs*. Trans. by Ott, Michaela. Zürich / Berlin: Diaphanes.

Garber, Daniel (1995): *Leibniz: Physics and philosophy*. In: Jolley, Nicholas (ed.): The Cambridge Companion to Leibniz. New York / Melbourne: Cambridge University Press.

Herder, Johann Gottfried (1778a): *Plastik. Einige Wahrnehmungen über Form und Gestalt aus Pygmalions bildendem Traume*. In: id.: Herders Sämmtliche Werke, vol. 8. Ed. by Suphan, Bernhard. Berlin: Weidmannsche Buchhandlung 1892, pp. 1–87.

Herder, Johann Gottfried (1778b): *Sculpture. Some observations on Shape and Form from Pygmalions Creative Dream*. Trans. and ed. by Gaiger, Jason. Chicago / London: The University of Chicago Press 2002.

Kafka, Franz (1948): *The Diaries of Franz Kafka*. Ed. by Brod, Max. Trans. by Kresh, Joseph. New York: Schocken.

Klee, Paul (1953): *Pedagogical Sketchbook*. Trans. by Moholy-Nagy, Sibyl. New York: Frederick A. Praeger.

Klee, Paul (1956): *Das bildnerische Denken. Schriften zur Form- und Gestaltungslehre*. Ed. by Spiller, Jürg. Basel / Stuttgart: Schwabe.

Lærke, Mogens (2015): *Five Figures of Folding: Deleuze on Leibniz's Monadological Metaphysics*. In: British Journal for the History of Philosophy, vol. 23, no. 6, pp. 1192–1213.

Leibniz, Gottfried Wilhelm (1676): *Pacidius an Philalethes. Grundphilosophie der Bewegung*. In: id.: Schöpferische Vernunft. Schriften aus den Jahren 1668–1686. Trans. by von Engelhardt, Wolf. Münster / Cologne: Böhlau 1995, pp. 102–168.

Leibniz, Gottfried Wilhelm (1849–63): *Mathematische Schriften von Gottfried Wilhelm Leibniz*, 7 vols. Ed. by Gerhardt, Carl Immanuel. Berlin: A. Asher / Halle: H.W. Schmidt.

Leibniz, Gottfried Wilhelm (1965a): *Die philosophischen Schriften von Gottfried Wilhelm Leibniz*. Ed. by Gerhardt, Carl Immanuel. Hildesheim: Georg Olms Verlag.

Leibniz, Gottfried Willhelm (1965b): *Monadology and Other Philosophical Essays*. Trans. by Schrecker, Paul / Schrecker, Anne Martin. Indianapolis: Bobbs-Merrill.

Leibniz, Gottfried Wilhelm (1985): *Theodicy: Essays on the Goodness of God, the Freedom of Man and the Origin of Evil*. Trans. by Huggard, E.M.. La Salle, IL: Open Court.

Leibniz, Gottfried Willhelm (1989): *Philosophical Essays*. Trans. and ed. by Ariew, Roger / Garber, Daniel. Indianapolis: Hackett.

Leibniz, Gottfried Wilhelm (1996): *Neue Abhandlungen über den menschlichen Verstand* (Philosophische Werke: in vier Bänden, vol. 3). Trans. and ed. by Cassirer, Ernst. Hamburg: Felix Meiner Verlag.

Leibniz, Gottfried Wilhelm (1999): *Sämtliche Schriften und Briefe*. Akademie Verlag: Berlin.

Leibniz, Gottfried Wilhelm (2001): *The Labyrinth of the Continuum. Writings on the Continuum Problem, 1672 – 1686*. Trans. and ed. by Arthur, Richard W. T.. New Haven: Yale University Press.

Leibniz, Gottfried Wilhelm (2002): *Monadologie und andere metaphysische Schriften. Discourse de métaphysique. La monadologie. Prinicpes de la nature et de la grâce fondés en raison*. Ed. and trans. by Schneider, Ulrich J.. Hamburg: Felix Meiner Verlag.

Leibniz, Gottfried Wilhelm (2008): *Protogaea*. Trans. and ed. by Cohen, Claudine / Wakefield, Andre. Chicago / London: The University of Chicago Press.

Lucretius (1994 – 2000): *On the Nature of Things*. Trans. by Leonard, William E.. Online: *http: //classics. mit.edu / Carus / nature_things.html* (last accessed 10 December, 2015).

McDonough, Jeffrey K. (2014): *Leibniz's Philosophy of Physics*. In: Zalta, Edward N. (ed.): Stanford Encyclopedia of Philosophy. Online: *http: //plato.stanford. edu / archives / spr2014 / entries / leibniz-physics* (last accessed 10 December, 2015).

Mercer, Christia / Sleigh, Jr., Robert C. (1994): *Metaphysics: The early period to the Discourse on Metaphysics*. In: Jolley, Nicholas (ed.): The Cambridge Companion to Leibniz. New York / Melbourne: Cambridge University Press, pp. 67 – 123.

Moser, Jakob (2014): *DAEDALA LINGUA. Lukrez als Übersetzer des Realen*. Online: *https: //www. academia.edu / 7429466 / Daedala_lingua_Lukrez_ als_Übersetzer_des_Realen_2014* (last accessed 10 December, 2015).

Mutsaers, Charlotte (1997): *Rachels Röckchen*. Trans. by Müller-Haas, Marlene. München: Hanser Verlag.

Nietzsche, Friedrich (1882): *The gay science. With a prelude in Rhimes and an appendix of songs*. Trans. by Kaufmann, Walter. New York / Toronto: Random House 1974.

Rutherford, Donald (1994): *Metaphysics: The late period*. In: Jolley, Nicholas (ed.): The Cambridge Companion to Leibniz. New York / Melbourne: Cambridge University Press, pp. 124 – 175.

Uhlig, Ingo (2007): *Poetologien der Abstraktion: Paul Klee, Gilles Deleuze* In: Blümle, Claudia / Schäfer, Armin (eds.): Struktur, Figur, Kontur. Abstraktion in Kunst und Lebenswissenschaft. Zürich / Berlin: Diaphanes, pp. 299 – 316.

Warburg, Aby M. (1893): *Sandro Botticellis Geburt der Venus und Frühling*. In: id.: Ausgewählte Schriften und Würdigungen. Ed. by Wuttke, Dieter. Baden-Baden: Koerner 1979, pp. 11 – 63.

Whitehead, Alfred N. (1978): *Process and Reality. An Essay in Cosmology* (Gifford Lectures, 1927 – 28). Ed. by Griffin, David R. / Sherburne, Donald W.. New York: Free Press.

Angelika Seppi
Email: *angelika.seppi@hu-berlin.de*

Image Knowledge Gestaltung. An Interdisciplinary Laboratory.
Cluster of Excellence Humboldt-Universität zu Berlin.
Sophienstrasse 22a, 10178 Berlin, Germany.

Claudia Blümle

Infinite Folds: El Greco and Deleuze's Operative Function of the Fold

The diffuse light and the vaporous masses of clouds bursting into the geometrically constructed space in El Greco's *Annunciation* of 1567 emphasise the contrast between a measurable terrestrial world and an immeasurable celestial one (fig. 1). The geometric shapes on the floor, with their closed contours, stand out against the billowing cloud formations, which carry the archangel Gabriel and also conceal most of the blue sky and the source of light. The opposition and transition between fixed forms and formlessness is carefully composed: the spaces occupied by the terrestrial floor and the heavens each take up a complete half of the painting.[1] Located precisely in the centre, the dark brown horizontal parapet, both horizon and demarcation of the floor, marks the separation between the two worlds.

Concealed behind the clouds, a wide, unbounded sky opens up above the horizon. Filling the upper part of the painting, it is either a celestial irruption into an interior, or a landscape that is located outside. The horizon above the parapet envelops the viewer as it would in a landscape painting. Both forms of the ground, the one of the floor and that of the landscape, are thus intertwined in El Greco's *Annunciation*.[2] There is a tension between the closed contours, the fixed forms and perspectival geometry of the lower half and the formless fields of colour, the tangents, curves, hyperbolae and undulating lines above that exceed the painting. This tension, a continuous motion, is created, as will be shown in what follows, by the different forms and processes of the folds in the painting.

---

1   Wölfflin 2015.
2   In his essay 'L'art et le pouvoir du fond' Maldiney shows that in the most powerful artworks we perceive the ground as analogue to the ground on which we move, or to the space that surrounds us and that we pass through; Maldiney 1994b, 182.

Fig. 1: Doménikos Theotókopoulos, called El Greco: *Annunciation*, about 1576, oil on canvas, 117 x 98 cm, Madrid, Museo Thyssen-Bornemisza

In El Greco's *Annunciation*, the chequered floor represents a tangible, clearly delineated space, in which Mary and the archangel Gabriel can take their positions. Like a wall, the low parapet, functioning as a border, marks the outline of the floor. The edge of the dark brown, impermeable wall acts as a distant horizon, on which we can discern the silhouette of a building. The vanishing point is thus situated behind the enclosing parapet, in a place which also lies outside our perception (fig. 2). The wall is not only a horizon and a dividing line between heaven and earth, but it also conceals, like a closed door, the distant recession of the floor as it converges into the infinite vanishing point. The

Fig. 2: Doménikos Theotókopoulos, called El Greco: *Annunciation*, about 1576, oil on canvas, 117 x 98 cm, Madrid, Museo Thyssen-Bornemisza, according to a scheme of C. B.

Fig. 3: Doménikos Theotókopoulos, called El Greco: *Annunciation*, about 1576, oil on canvas, 117 x 98 cm, Madrid, Museo Thyssen-Bornemisza, detail.

position of the bright cloud in front of the massive, dark and enclosing wall intensifies the downward pull of the linear perspective. In this place in the painting, the encounter of the terrestrial world and the numinous realm, the divine, is palpable: on the floor we can see the distinct outlines of the shadow cast by a cloud (fig. 3). In its fusion of light and line, the representation of this shadow, with its closed contours, follows the rules of linear perspective, while the indeterminate form of the cloud itself defies such a spatial representation.[3] In opposition to the pure perspectival clarity, the opaque cloud, which is both light and heavy, bright and dark, almost completely obscures the view of the wall (fig. 1).

Six years before the *Annunciation* of 1576 (fig. 1), which measures 117 x 98 cm, El Greco painted a smaller version of the same theme (fig. 4). As in the later version, bright light, accompanied by tumbling cherubs, pours out of the cloudy sky, streaming down into the geometrical space below where it seizes the red curtain. In both paintings, the opposition of solid form and diffuse formlessness corresponds to the terrestrial world below and the celestial sphere above, a physical and a numinous world. It is within this field of tension that the Annunciation takes place, and El Greco's task was to render visible this biblical event of the Incarnation as transition. As in the later treatment, in this earlier, smaller version, a linear perspective pulls the viewer's gaze to a distant place. With its steep recession, the painting establishes a rhythm alternating between the light falling in from the right and the shadow cast by the cloud. If we retrace the thinly drawn grid of the floor, we find two vanishing points. The location of the first – in the shadow of the larger arch, where the two central orthogonals intersect – is very close to the threshold of inside and outside. This disturbs the spatial arrangement: the view of the outside appears as a painting within the painting. But if we follow the outer orthogonals, we arrive at a second vanishing point, in the arch at the end of the narrow passage (fig. 4). The orthogonals do not end behind an opaque wall here but recede into a view opened up by the arch. The patch of blue within this second arch either suggests a view on to an open sky or a marble wall. El Greco's two *Annunciations* show how he developed his pictorial concepts of form and linear perspective, of opaque formlessness and deformation, and brought them into a dynamic relationship to each other.

In his study *Francis Bacon: The Logic of Sensation*, first published in 1981, Gilles Deleuze has explored the deformation of painting in El Greco's work. He draws our attention to a horizontal line in *The Burial of the Count of Orgaz*, which divides the painting into two parts, "upper and lower, celestial and terrestrial"[4], a division which we also find in El Greco's *Annunciation*. In the lower part

> "there is indeed a figuration or narration [...], although all the coef-
> ficients of bodily deformation, and notably elongation, are already
> at work. But in the upper half [...] there is a wild liberation, a total
> emancipation: the Figures [...] are relieved of their representative role,
> and enter directly into relation with an order of celestial sensations."[5]

---

3   Brunelleschi demonstrated this in his experiment with linear perspective in the early fifteenth century.
     See Damisch 2002, 111–124.
4   Deleuze 2003, 9.
5   Ibid.

Fig. 4: Doménikos Theotókopoulos, called El Greco: *Annunciation*, about 1570, oil on wood, 26 x 20 cm, Madrid, Prado.

With the Holy Ghost, "lines, colors, and movements are freed from the demands of representation."[6] Seven years later, in his book *Le pli* (*The Fold*) of 1988, Deleuze returns to this horizontal line which divides the painting in two, creating an upward surge and downward pull. Deleuze considers this fold, "a fold which differentiates and self-differentiates"[7], to be the ideal fold of the baroque, and adopting a Heideggerian concept, calls it *Zwiefalt*. In El Greco's *Annunciation* this "twifold"[8], a fold between two folds, articulates itself on the horizontal and vertical planes, in curvilinear and undulated form and in the fields of colour.

According to Deleuze, a painted image can be conceptualised in two ways: the Platonic notion of everlasting similitude and identity or exactly the opposite, with difference at its core, and representations as the product of a movement of difference. Deleuze challenges the notion of a painting as static representation, depiction, narrative or illustration of something external to itself, arguing that it is the difference internal to the painting that unfolds as creative movement, revealing a loose and multidirectional network of colours

6   Ibid.

7   Deleuze 1991, 236.

8   In the English edition of *Le pli* (Deleuze 1993) Heidegger's neologism *Zwiefalt* has been erroneously rendered as *Zweifalt*, which, as Sjoerd van Tuinen points out, does not capture the word's intrinsic duplicity and ambiguity. He suggests that the best English translation would be the Middle English "twifold". Van Tuinen 2014, 69 and 83f. (note 49).

and lines. The static opposition of colour and ground is dissolved and replaced by a fluid interplay that has emerged from dynamic structural relations.[9] In El Greco's *Annunciations* the formless ground of colour and the clearly outlined figures have equal weight, and their relation to each other is one of difference as well as of gradual transition. The motifs of heaven and curtain are also spatially separate yet intertwined, and this interpenetration is most intense where the celestial light dissolves into the drapery of the red curtain. The opposition of the numinous-intelligible world and the physical world of sensations is overcome through the curtain as well as the mediation of the archangel Gabriel, who "is the figure of the coming of the invisible into the visible and of the unsayable into the word"[10]. That he acts as a messenger is emphasised by his central place in the composition, opposite the red curtain, exactly between heaven and earth. Gabriel's arrival on earth is marked by the sharply defined shadow on the one hand and the formless cloud on the other. The form of the archangel, who is floating above the ground in his heavenly sphere, with a bared leg, is gracefully distorted into a *figura serpentinata*.

Fig. 5: Doménikos Theotókopoulos, called El Greco: *Annunciation*, about 1576, oil on canvas, 117 x 98 cm, Madrid, Museo Thyssen-Bornemisza, according to a scheme of C. B.

In direct contrast, the straight vertical line of his sceptre points towards the heavens and intersects the horizontal line of the wall (fig. 5). If we follow the angel's downturned right wing, we notice that it describes a horizontal line, while the left wing, gradually turning from white to black, ascends into the cloudy sky as a vertical line parallel to the frame. A structure of perpendicular lines is thus repeated in metonymic shifts in the

---

9   Uhlig 2007, 301–305.
10   Marin 2006, 181.

painting. Even the smallest details have this basic structure of a cross, for example the filial of the sceptre with its three white pearls, which marks a horizontal to both sides and also joins the ascending vertical. Thus two movements are created: a vertical ascension and a horizontal bilateral movement, which points sideways out of the picture and is echoed on both sides by the two figures, Mary and the archangel Gabriel, as well as by the swathes of red fabric next to them, the curtain to the left and the drapery beside the angel on the right.

Rich in tonal variation, the folds of the red curtain, which reach into the light and colour of the heavens, folding and unfolding from both sides and culminating in *Zwiefalt*, move into the darkness at the left margin and the radiant light in the distance. This folding and unfolding movement breaks out of the frame as an infinite sequence, abandoning the geometry of the finite for the differential geometry of calculus. In the first two volumes of his *Konturen einer Geistesgeschichte der Mathematik: Die Mathematik und die Wissenschaft* (vol. I) and *Die Mathematik in der Kunst* (vol. II)[11] Max Bense refers to Heinrich Wölfflin's analysis of form[12], which, along with the work of Alois Riegel and Wilhelm Worringer, was crucial for Deleuze's conceptualisation of painting. Bense theorises the relationship between the principles of style as form in art and in mathematics to explore parallel developments between the history of art and the history of mathematics. According to Bense, Wölfflin's studies of the transition from Renaissance to baroque as a movement from fixed form to formlessness[13], from the linear to the painterly, from planarity to recession, closed form to open form or clearness to unclearness[14] reveal formal developments in painting that attend Euclidean geometry in the Renaissance and elements of calculus in the baroque. The baroque handling of light and colour and its curved lines can be expressed mathematically as infinitesimal functions.[15] In its beauty "of boundlessness and the infinite"[16], the open form of the baroque emancipates itself from the frame, filling it with unruly figures, unevenly distributed areas of light and sharply cropped forms.[17] By using terms such as *sequence*[18], *irrational*[19] and *incommensurable*[20] for the description of the dynamic depth of baroque paintings, Wölfflin, for his part, applies mathematical concepts to his analysis of form.[21]

---

11   Bense 1948 and Bense 1949. For the link between Bense and Wölfflin see: von Herrmann 2014.
12   Wölfflin 1967 and Wölfflin 2015.
13   Wölfflin 1967 and Wölfflin 2015.
14   Wölfflin 2015.
15   Bense 1948, 26f. and 92.
16   Wölfflin 2015, 226.
17   Ibid, 205 and 210f.
18   Ibid, 230.
19   Ibid, 278, 280, 283, 291.
20   Ibid, 291.
21   Bense 1948, 100.

Art-historical, philosophical and mathematical analysis of form have helped us to elucidate the structure of El Greco's *Annunciation*: the horizon line, which demarcates zones of drawn form and coloured formlessness, the geometrical perspectival space or the intersecting vertical and horizontal movements above the archangel. Further connections to calculus can be seen in one of the important figures in the composition, the red curtain, which takes up two thirds of the pictorial space on the left side. Falling down in a straight line in one place and in undulating folds in another, it creates a vertical line which joins the figure of Mary. What is striking here is that the edge of Mary's blue garment is aligned exactly with the horizontal line of the grid of the floor. If we read this horizontal line as the axis of a coordinate system and take the perpendicular line of the left edge of the painting as vertical axis, we can retrace the figure of a hyperbola, which escapes the frame of the painting (fig. 6). The complex structure of this work thus shows baroque space curves in the interplay of coordinate axes and tangents. Bense explains this process in baroque art as follows:

> "The simple, continuous linear arrangement of objects commonly
> used in Renaissance painting to create clear, simple ratios of pro-
> portion and symmetry, which, as Wölfflin too has rightly noted, also
> governs all classical art, is replaced in the baroque by a non-linear,
> continuous, curved arrangement of objects, and from this results
> what some describe as the emphatically asymmetrical character of
> baroque composition." [22]

The pictorial and compositional hyperbola of the red curtain at the left margin in El Greco's *Annunciation* and the superimposed undulating line of the drapery further unfold in the folds that merge into the heavens, fan out of them or are drawn in. The interaction between the cloud formations and light trails on one side and the folds of the curtain and its shadow on the other is also expressed as an upward-sloping curve under the group of cherubs.

The undulating lines, curves, zigzags and diagonals are not drawn, but created by the interplay of colour modulations between light and dark, between warmer and colder tones. This complexity of baroque drapery as *Zwiefalt* did not appear in the earlier treatment of 1560. Here the curtain moves in a diagonal line, merging into the light and reaching

---

22   Bense 1949, 58. Note that this does not refer to the 17th century Cartesian coordinate system of analytic geometry (developed by Descartes and Fermat), which might have also oblique axes, but to a late Middle Ages concept of geometry that emphasizes the construction of an infinite space via a system of perpendicular lines. For concepts of an infinite, imaginary space (*spatium imaginarium*) during the late Middle Ages see Grant 1981, chapter 6. Bense has also suggested that El Greco's paintings served Leibniz as a source of inspiration for the discovery of calculus. Note that calculus was developed by Gottfried Leibniz and Isaac Newton, independently from each other, in the 17th century.

Fig. 6: Doménikos Theotókopoulos, called El Greco: *Annunciation*, about 1576, oil on canvas, 117 x 98 cm, Madrid, Museo Thyssen-Bornemisza, according to a scheme of C. B.

out of it into the pictorial space. But a space curve, framed by the white stairhead, is perceptible as a contour in Mary's garment. In both treatments of the *Annunciation* the finite in the geometrical space is pushed to its limits, both undermined and mediated, to create a bridge to infinitude and formlessness: a concentric movement penetrates the depth of the painting, while an eccentric movement reaches out of it. While the movement of the systole leads to an isolation of the figure, that of the diastole causes expansion and dissipation, as Deleuze has shown in *Francis Bacon: the Logic of Sensation*.[23] Within the painting everything is distributed as diastole and systole. The systole contracts and condenses forms and bodies, while the diastole expands and dissolves them into light and colour. Even if the form of the curtain were to dissipate entirely in formlessness, it would remain constrained by the forces that have taken hold of it to return it to its surroundings. Deleuze calls the interplay of diastolic and systolic movements *rhythm*. [24] It is through rhythm that form manifests itself in the painting and the painting's relation to the outside is expressed.

The relational web in this painting follows Deleuze's concept of the diagram as a disorganised and deformed unity, which shows a visual abyss in the depths but also creates order and rhythm. The diagram brings violent chaos and formless colour to the pre-existing figuration but is also "a germ of rhythm"[25] for the new order of painting. A year

23 Deleuze 2003.
24 Ibid.
25 Ibid, 102.

before *Francis Bacon: The Logic of Sensation*, Gilles Deleuze and Félix Guattari explored rhythm and its relation to chaos. "From chaos, *Milieus* and *Rhythms* are born." Chaos is the milieu of all milieus. The milieus that have come out of chaos either consolidate or revert, dissolving back into chaos. In the milieu as form, chaos is never absent. Rhythm is created through passages from one milieu to another.[26] The in-between is what rhythm and milieu have in common, while the emergence of rhythm between the milieus that have rhythm creates difference. This is why rhythm and milieus that have rhythm are never on the same plane.[27]

The consolidating and closed milieus, which are deformed in El Greco's paintings, play a crucial role for Deleuze and Guattari and their approach to art. Art prevents milieus from stabilising, enabling instead an open and developing in-between of rhythm. In their philosophy of becoming, Deleuze and Guattari see rhythm as a movement between chaos as ground and emerging milieus. What is created by the operative identity of the fold is neither a static form nor a spatial juxtaposition of forms but the rhythm in its space-time dimension.

In El Greco's *Annunciation*, the divine bursts into the world through the medium of painting: the Incarnation as the becoming of God emerges from the shapeless ground as colour. Out of the colour of the heavens, the Holy Ghost penetrates the room, followed by the archangel Gabriel, who surprises Mary as she is reading a book. From behind the dark cloud formations in the distant depths of the painting, the Holy Ghost, in the shape of a white dove, emerges out of the radiant yellow ground. The archangel Gabriel is enclosed by grey-brown clouds and blue sky, and Mary stands out against the red curtain. While these figures are clearly delineated from their surroundings, the celestial colours on the left-hand upper margin of the painting all merge into each other. The yellow, blue, grey-brown and red are the ground of the source of light, of the sky, the clouds and the curtain. They create metonymic movements in the painting, which enable an interaction between the formless ground and the solid forms of the figures. The yellow of the ground behind the clouds is echoed in Gabriel's garment, the blue of the sky on the right reappears in Mary's cloak, and the red of the curtain in the curling drapery, which ends in a circular form in the far right of the painting.

Not form but colour determines the ground of the numinous divine, which streams into the geometrically constructed space. Like a magnetic force field, the celestial ground seizes the red curtain. At its centre are the fields of colour – the grey-blue of the sky and the yellow behind the clouds – and the Holy Ghost and the cherubs. The fiery golden yellow pouring out of the depths behind the parting clouds, carrying the Holy Ghost

26   Deleuze/Guattari 1988, 313.
27   Ibid.

in the form of a white dove, is particularly intense and powerful. This yellow breach in the clouds is the pivotal point of the painting. The diagonal between the archangel's raised arm and a floating cherub crosses the trail of light directed towards Mary, which also flows into the right wing of the upper cherub. The archangel Gabriel gazes down at Mary and transmits the divine message, pointing towards the centre of the painting, the yellow ground.

Thus in the *Annunciation* of 1576 the archangel's pointing finger directs us back to the location of the vanishing point. If we ask what is hidden behind the billowing cloud formation, our attention is drawn to the pictorial layers unfolding in the distance. As can be seen from the shadow cast by the cloud, the archangel Gabriel is standing behind Mary. Mary is therefore in the foreground of the painting, but her body is twisted backwards, so that she can see the archangel. The archangel looks at and points towards the Holy Ghost in the form of the white dove. The movement into the distant background, as well as into the horizontal width and vertical height, opens up an almost unfathomable depth. This pictorial depth, which sets in motion a play of ever more distant vistas, posing the ontological and theological question of the ground as ground of being, is opened by the folds of the red curtain. Abandoning the spatial staggering of the Euclidean linear perspective, still present in the tiled floor and the wall, the opaque curtain emerges as a field of colour that renders visible a movement between form and formlessness. Folded as *Zwiefalt* it mediates between the perspectival space and the radiant celestial realm behind the clouds. It thus establishes the revealing and concealing function of the ground, which opens up the unfolding of sensation.

The folding and unfolding between form and formlessness enters the painting through the gazes and the constellation of gazes of the cherubs. The cherubs accompany the flight of the dove, and the manner of their depiction makes the Holy Ghost appear even more distant in the depth of the infinite ground. In both *Annunciations*, the entrance of the Holy Ghost is expressed in the form of light and shade, as well as in the drapery of the red curtain. In the 1570 version, which also contains a small red curtain, the warm light surrounding the Holy Ghost is in stark contrast to the cold light outside the room, highlighting the irruption of the divine into the terrestrial space. These movements in the painting help to emphasise Mary's backward turn towards the archangel who is standing behind her. The latter points towards the Holy Ghost and the radiant yellow ground, which initiates a back-and-forth and circular motion between the near and far in the painting. While in the 1570 version the attitudes of the cherubs directed this movement into and out of the remote depths of the painting towards the near (fig. 2), in the 1576 version the movement runs across the whole painting through the constellation of gazes (fig. 1). Mary's turning towards the archangel, who is standing behind her, is reciprocated by Gabriel's eye contact, as he points towards the Holy Ghost and the red curtain. At the same time, the three cherubs at the top left look at Mary. If we follow the directions of the gazes of all the figures in the painting, we see that they describe a circle, which corresponds to the spatial movement into the distance, towards the radiant ground behind the clouds and back towards the near foreground, to Mary (fig. 7). This movement, a vector field which exists only on the surface, intersects the geometric planes of the composition. [28] The diagram, as operative set of asignifying lines and fields of colour, moves between the fluid chaos of formlessness and the static order of form.

Fig. 7: Doménikos Theotókopoulos, called El Greco: *Annunciation*, about 1576, oil on canvas, 117 x 98 cm, Madrid, Museo Thyssen-Bornemisza, according to a scheme of C. B.

The movement from the foreground to the deepest layer of the radiant yellow ground is also reversed: the light, which emanates from the lowest layer of the divine ground, penetrates the room, describing a movement from the background into the foreground. All the lighting in the painting is arranged around the source of light behind the Holy Ghost. This becomes especially obvious in the tonal modulations of the archangel's right lower arm and his neck, as well as Mary's hands, face and neck, and the bodies of the small cherubs in the upper section of the painting, or the curtain on the left. This countermovement is emphasised by the three cherubs above the yellow ground, who look down towards the red curtain. The light bursts out of the yellow ground that seems to be located behind the thick, dark blue clouds. But if we look at the zone underneath we see that the yellow light rays are now appearing in front of the cloud formations rather than behind them. In this place the ground, from which the light emanates, has turned into rays of yellow light, which penetrate the room and create the shadow of the floating cloud.

---

28   I am grateful to Michael Friedman for this observation.

In both *Annunciations* by El Greco, the use of light and shade clearly differentiates active and passive ground, a point that Deleuze, referring to *The Baptism of Christ*, has commented on in *The Fold*. El Greco's paintings show that the darkness and the light are both active grounds.[29] The yellow ground, concealed behind the dark clouds but breaking out of them in the form of the Holy Ghost, is an active ground, which takes hold of the forms and deforms them. What is crucial here is the colour, as well as the circular relation between clearly formed and formless ground, a relation that manifests itself in the restless twists and turns of the cherubs, until it finally streams into the room as incandescent light. In the 1570 *Annunciation,* too, the distant yellow ground articulates itself, breaking through the billowing dark clouds. In both treatments the red curtain is seized by the active force of the yellow ground from which the light source of the painting streams. It simultaneously dissolves in it and emanates from it, folding and unfolding.

El Greco's paintings direct our attention to the non-visible, to the formless and the fields of colour, which in their visibility make the sensible experienceable as withdrawal. It is not recognisable objects and given forms referring to an intelligible world that take central stage in the *Annunciations* but rather the folding and unfolding of an infinite becoming. God is being itself and the ground of Being, and God's incarnation raises the question of the ground of images, which is why Louis Marin could conclude that the Incarnation concerns the theological essence of painting itself.[30] Representing the moment of Incarnation, both *Annunciations* by El Greco show that the essence of the ground lies in its continuous withdrawal. But this withdrawal does not refer to the invisible world of ideas, which primarily shapes sensation, but marks the beginning of visual emergence. Not as an intelligible sign, but as a shapeless colour, the divine ground breaks into the visible world. The mystery of the Incarnation is translated into the process of artistic creation, appearing in the painting as an infinite and continuous movement of the fold between form and ground.

Drawing on Henri Maldiney's essay 'L'esthétique des rhythmes',[31] Deleuze has related this movement and the resulting deformation to the baroque fold, "which ceaselessly unfolds and folds back from both sides and which only unfolds one by folding back the other in a coextensivity of the unveiling and veiling of Being, of the presence and withdrawal of the being." [32] What is shown in this concept of the image is thus the boundless freedom of

---

29 Max Raphael has shown this in a comparison of Tintoretto and El Greco. See Raphael 2009, 111.

30 Marin 2006, 195.

31 See Maldiney 1994a. Drawing on F. W. J. Schelling's concept of the *Grund* (ground), Henri Maldiney stresses that the ground is the foundation that enables the forms to liberate themselves from the surroundings that have captured them. But the German word *Grund*, as he goes on to explain, also means origin and cause. Thus the ground does not only function as counterpart of the forms but also establishes an ontological and aesthetic relation to the world. See also Maldiney 1994b, 174.

32 Deleuze 1991, 236.

the fold. As the fold frees itself from its supports in the painting, an infinite competition between the structure of the fold and the objects begins, which Deleuze demonstrates using two examples from the work of El Greco:

> "If we wish to maintain the operative identity of the Baroque and the fold, we must then show that in all other cases the fold remains limited while in the Baroque it experiences a limitless release, whose conditions can be determined. The folds seem to take leave of their supports, cloth, granite, and cloud, to enter into an infinite competition, as in the *Christ in the Garden of Gethsemane* of El Greco (the one in the National Gallery). Or else, notably in *The Baptism of Christ*, the counter-fold of calf and knee, where the knee seems the inversion of the calf, lends an infinite undulation to the leg, while the pinching of the cloud in the center transforms it into a double fan ..."[33]

The counter-fold, the infinite undulating form and the double fans are rendered visible in the two *Annunciations*, as the fold leaves its supports – curtain, draped garments and clouds – and, as a detached and free structure, is transferred to the painting as a whole, where it captures the field of the Elements:

> "The liberation of folds that are no longer merely reproducing the finite body is easily explained: a third entity, or entities, have placed themselves between clothing and the body. These are the Elements. We need not recall that water and its rivers, air and its clouds, earth and its caverns, and light and its fires are themselves infinite folds, as El Greco's paintings demonstrate."[34]

In El Greco's two *Annunciations*, the elements "mediate, distend and broaden" not only the clothing and the bodies but also the drapery of the red curtain with its deep shadows. [35] At first, the drapery, with its dark modulations of shade, appears in spatial mode. But at the moment when the curtain is seized by the pictorial ground of air, light, fire and clouds, its operative identity also becomes visible in temporal mode, opening up a communication between form and formlessness, near and far, inside and outside, the finite and the infinite.

Translation: Martina Dervis (London)

---

33   Ibid, 241f.
34   Deleuze 1993, 122, trans. modified.
35   Ibid.

Bibliography

Bense, Max (1948): *Konturen einer Geistesgeschichte der Mathematik, vol. I: Die Mathematik und die Wissenschaft*. Hamburg: Claassen & Goverts.

Bense, Max (1949): *Konturen einer Geistesgeschichte der Mathematik, vol. II: Die Mathematik in der Kunst*. Hamburg: Claassen & Goverts.

Damisch, Hubert (2002): *A Theory of / Cloud /: Toward a History of Painting*. Trans. by Lloyd, Janet. Stanford, CA: Stanford University Press.

Deleuze, Gilles (1991): *The Fold*. Trans. by Strauss, Jonathan. In: Yale French Studies, no. 80, Baroque Topographies: Literature / History / Philosophy, pp. 227–247.

Deleuze, Gilles (1993): *The Fold: Leibniz and the Baroque*. Trans. by Conley, Tom. London: The Athlone Press.

Deleuze, Gilles (2003): *Francis Bacon: the logic of sensation*. Trans. by Smith, Daniel W.. London / New York: Continuum.

Deleuze, Gilles / Guattari, Félix (1988): *A Thousand Plateaus: Capitalism and Schizophrenia*. Trans. by Massumi, Brian. London: The Athlone Press.

Grant, Edward (1981): *Much Ado about Nothing: Theories of Space and Vacuum from the Middle Ages to the Scientific Revolution*. Cambridge: Cambridge University Press.

Maldiney, Henri (1994a): *L'esthétique des rhythmes*. In: id.: Regard, Parole, Espace. Lausanne: Editions l'Age d'Homme, pp. 147–172.

Maldiney, Henri (1994b): *L'art et le pouvoir du fond*. In: id.: Regard, Parole, Espace. Lausanne: Editions l'Age d'Homme, pp. 173–207.

Marin, Louis (2006): *Opacité de la peinture: Essais sur la représentation au Quattrocento*. Paris: Editions EHESS.

Raphael, Max (2009): *El Greco: Ekstase und Transzendenz. Mit Bildvergleichen zu Tintoretto*. Ed. by Heinrichs, Hans-Jürgen. Berlin: Reimer.

van Tuinen, Sjoerd (2014): *Difference and Speculation: Heidegger, Meillassoux, and Deleuze*. In: Beaulieu, Alain / Kazarin, Edward / Sushytska, Julia (eds.): Gilles Deleuze and Metaphysics. London: Lexington Books, pp. 63–90.

von Herrmann, Hans-Christian (2014): *Dämonie der Technik. Max Benses Geistesgeschichte der Mathematik*. In: Friedrich, Lars / Geulen, Eva / Wetters, Kirk (eds.): Das Dämonische. Schicksale einer Kategorie der Zweideutigkeit nach Goethe. Paderborn: Wilhelm Fink, 2014, pp. 363–372.

Uhlig, Ingo (2007): *Poetologien der Abstraktion: Paul Klee, Gilles Deleuze*. In: Blümle, Claudia / Schäfer, Armin (eds.): Struktur, Figur, Kontur: Abstraktion in Kunst und Lebenswissenschaften. Berlin / Zurich: Diaphanes, pp. 299–316.

Wölfflin, Heinrich (1967): *Renaissance and Baroque*. Trans. by Simon, Kathrin. Ithaca, NY: Cornell University Press.

Wölfflin, Heinrich (2015): *Principles of Art History: The Problem of the Development of Style in Early Modern Art*. Trans. by Blower, Jonathan. Los Angeles, CA: Getty Research Institute.

Claudia Blümle
Email: *claudia.bluemle@hu-berlin.de*

Image Knowledge Gestaltung. An Interdisciplinary Laboratory.
Cluster of Excellence Humboldt-Universität zu Berlin.
Sophienstrasse 22a, 10178 Berlin, Germany.

Emmanuel Ferrand, Dominique Peysson

Versal Unfolding:
How a Specific Folding Can Turn Crease
and Tear into Transversal Notions

Preamble

> "L'enfant naît avec vingt-deux plis. Il s'agit de les déplier. La vie de l'homme
> est alors complète. Sous cette forme il meurt. Il ne lui reste aucun pli
> à défaire."
> "The child is born with twenty-two folds. What needs to be done is to
> undo the folds. Then the life of the man will be complete. This is the
> state in which he dies: nothing is left to unfold."

> Henri Michaux, *Au Pays de la magie*, 1941

> "A work of folding and unfolding in which every element becomes
> always the fold of another in a series that knows not point of rest."

> Stephen Heath, *Post-structuralist Joyce: Essays from the French*, 1984

From a mathematical perspective, folds and tears of paper may be considered dual notions: there is a strong relationship between the fold of a material and the expectation of a possible catastrophic tear. In this paper we recall a simple and concrete experiment that took the form of performance art. It revealed that, using only a few sheets of paper, while handling it carefully through folding, it is possible to avoid the pairing between folds and tears. To carry out its analysis, we will first review some aspects of the mathematical theory of singularities from René Thom's original viewpoint. Throughout its course we will touch on complex notions such as versal unfolding[1] and transversality[2] and complex texts – from James Joyce to Stéphane Mallarmé. Our conclusion will take the form of an additional performance to be shared with our willing reader.

This paper is concerned with a performance, which could be construed as a simple or naive form of scientific experiment in materials science. This experiment will be the starting point of a highly non-linear random walk through seemingly unrelated fields of science, literature and the arts. Our paper is not intended to be strictly scientific in essence or form, rather it should be regarded a performance in itself, one in which two authors, as two sides of a sheet of paper, share ideas that resonate through a process of folding and unfolding.

This paper is organized as follow. We begin with an account of a performance of ours, which took place at a workshop organized by the research group GDR Mephys at ESPCI. On that occasion we presented a folding process – one we refer to by a *paper junction* – which was invented for this purpose. It allows us to assemble a number of sheets of A4 paper to make a longer strip of paper. During the performance the robustness of this junction was tested in a way that involves the bodies of the two performing artists. We then follow with a motivation for our choice of performance. We recount additional experiments involving our paper junction, which we conducted in order to assess its robustness and its optimal usage. In the third part that follows, we gradually turn reading itself into a performative act, by folding in concepts of pure mathematics one after the other in the manner of accordion folds, in resonance with concepts that underlined our performance. We then continue with the fourth part of this paper, where we unfold and fold several sources that evoke the fold as a metaphor. We jointly discuss Joyce's *Finnegans Wake*, a possible principle of science creativity and Stéphane Mallarmé's poem *un coup de dé jamais n'abolira le hazard*, that epitomize our conception of the fold as what oscillates between metaphor and matter, between inside and outside, between the singular and the general. In the final part of this article, we therefore conclude with a performance that utilizes the very last figure in this paper – an actual folding act of the paper this article is printed on by readers themselves.

---

1   The notion of versal unfolding is a fundamental tool in the mathematical theory of singularities: a possibly singular object might be difficult to understand when considered *per se*, but is better understood when embedded in a family of often simpler configurations.

2   Transversality is an important notion in differential topology (Thom 1958). Two transverse submanifolds (e.g. curves drawn on a sphere or surfaces embedded in a three-dimensional euclidean space) intersect transversely if at every component of intersection (e.g. points resp. curves), their corresponding tangent spaces at that point generate the tangent space of the embedding manifold at that component.

1 A paper performance

Earlier this year we were invited to perform an art and science act titled *Onto the fold* as part of the workshop *Folding and Creasing of Thin Plate Structures* held at the ESPCI.[3] In what follows we will discuss this somewhat unusual contribution to a scientific conference.

Two performers get in front of an audience to prepare for what is about to happen. The experiment is quite simple. Its simplicity is demonstrated by making the process as straightforward and participatory as possible. Throughout the performance the weight of the two performers is used to test the strength of a paper junction designed especially for the performance. Its folding process can be seen in fig. 1. We refer to the resulting paper junction by the term *Peysson Junction*. The performance demonstrates how one can generate a firm join by folding together two sheets of standard A4 paper that combine through a series of steps to create a longer strip of paper. Joining together the two sheets does not involve any binding material (glue, thread, etc.) nor does it require any paper cutting. It only requires folding.

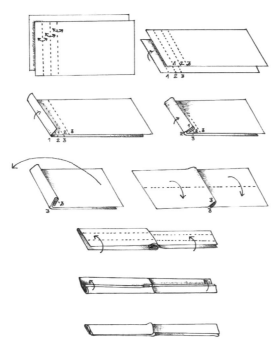

Fig. 1: Junction making guide

---

3   A one-day workshop organized by the research group GDR MePhy and Keith Seffen (Department of Engineering, University of Cambridge) held at the ESPCI, Paris, March 26, 2015. See Ferrand/Peysson 2015.

As the performance begins, four members of the audience are each handed a sheet of A4 paper. Each is asked to carry out a *J3 folding* defined as follows: folding comprises three equidistant folds (approximately 3 cm wide) that run parallel to the short edge of the paper. The four sheets of folded paper are then collected; two of them are now assembled according to the subsequent steps in fig. 1. Each two-sheet ensemble gets folded in half along the middle line that runs parallel the long edge. Each then gets evenly folded in three along two middle lines that run parallel the same long edge. We refer to the latter two steps of the sequence as a *P2-3 folding*. The resulting fold at this point is called a J3-P2-3 junction – none other than a Peysson junction.

Once each Peysson Junction is ready, force is applied to either end of the paper-strip, as each performer, grabbing hold of one end, starts leaning backwards. As the angle increases between the horizontal axis of the paper-strip and the axis of alignment of each performer's body, so does the intensity of the force applied at each end. The performers' bodies and the strip of paper together form a system, whose internal tension is gradually increasing. The experiment goes on up until its inevitable failure – one or both performers eventually fall to the ground.

There are a couple possible scenarios to explain the final rupture: either it occurs as the paper splits apart somewhere away from the fold, or it happens at the join. In the latter case, either the fold at the join comes undone as one sheet of paper slides off the other, or the paper splits apart along one or more of the fold-lines (as a result of a possible susceptibility induced by folding).

During the performance at the ESPCI it was the second author who took the fall. In that instance, the paper splint apart away from the join. It was the paper itself that proved weaker, as its fibers came apart. The folded junction on the other hand remained intact. The rupture ran perpendicular to the long axis of the paper-strip. The split was similar to what one would eventually expect to get, from a sustained simultaneous pull at two short edges of a single sheet of A4 paper. The result is shown in fig. 3.

2 A paper trail from the discrete to the continuous, and vice-versa

2.1 Foundations

The authors' paper performance has been designed with the intention of bringing together two pairs of concepts: 'discrete/continuous' as the first pair, and 'fold/tear' as the second. It differs from Origami, in that Origami typically engages a single sheet of paper, a fact, which renders a cohesion and resistance to the resulting object – a complex

Emmanuel Ferrand, Dominique Peysson

three-dimensional shape, which can possibly maintain some degree of flexibility. Either way, the result has a lower mechanical strength compared to that of the original sheet of paper (or sheets of paper), since folding introduces greater fragility.

Let us begin with discretization and continuity. Consider a standard sheet of A4 paper. Its overall area determines a unit of discretization – the discretization, for instance, of continuous text as it is set in print format.[4] Through our performance we examined the possibility of starting from a discrete set of objects – a few sheets of paper – so as to end with a design of a continuous object – a continuous strip composed of joined paper sheets. We wished to test whether we could generate good longitudinal resistance in such a way that weakness does result from juncture points. We wanted to see whether structure as a whole could be made ignorant of underlying discreteness. Our goal, through a process of trial and error, was to devise an overall folded structure to be much longer but at the same time as strong as the original sheet of paper itself, keeping in mind there is an obvious force threshold beyond which our paper strip will break, and our paper-system will get discretized again.

The dual pair fold-tear forms the second axis of our analysis. The fold is often used when one wants to rip (but not cut) paper along a certain line. A fold introduces a discontinuity in curvature to the otherwise flat, continuous surface of the sheet of paper. Energy, derived from torque applied to the planes on each side of the folded sheet of paper, is focused along the line of the fold – the line of flatness discontinuity.[5] Tear, as a result of ripping, at a given point along the ling of the fold, occurs when this energy exceeds a threshold. Ripping begins at the point where the fold line meets the border of the sheet of paper. A gradual separation into two parts then occurs. In this example, fold and tear appear as dual concepts: where a crease is, tearing will be.

2.2 Additional experiments

Our performance was supplemented by a series of resistance tests conducted in the laboratory so as to confirm the reproducibility of the art/science experiment. Five different tests were carried as follows (each repeated several times):

Test 1. Two sheets were joined together through a J3-P2-3 Peysson Junction (as during performance, see fig. 1). The resulting strip of paper was tested against sustained pull

---

4   Thus, for example, text flow is governed by the logic of content, but the setting of text into printed format requires dividing it in relation to a unit of area – the area of the sheet of paper in question. Compare section 4.3 in this paper.
5   The axis of torque runs perpendicular to the plane of the paper.

till it split apart. The remaining strip of paper – now shorter – was put to the test again till it split apart in its turn. Its remaining bit was then retested – now even shorter – till it broke apart as before. The process continued till the strip of paper got too short to be tested, nevertheless maintaining the J3-P2-3 join. The experiment was repeated three times from the beginning. Results were identical: at the end of each test the resulting tear ran perpendicular to the long axis of the strip, away from the junction, and rather close in fact to one or the other end (it was not possible to predict which end, though). See fig. 2.

Test 2. Three sheets were joined through two J3-P2-3 Peysson junctions. Same positive results as in the first test: the tear ran perpendicular to the long axis of the strip, away from the junction.

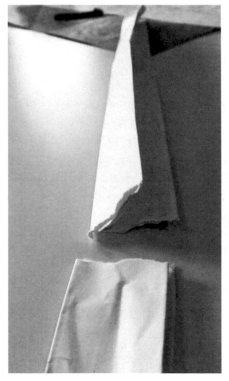

Fig. 2: The strip is torn away from the fold.

Test 3. Two sheets were joined through a *J2-P2-3 Junction*, namely, two folds at the first stage of folding instead of three. The results here were different. The strip of paper split apart along folds at the junction. All repeated tries of this experiments showed the same result: the fold ripped (fig. 3).

Fig. 3: Weaker junction.

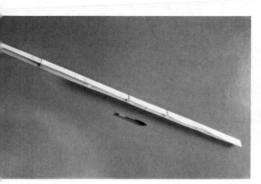

Fig. 4: A longer strip.

Test 4. Two sheets were joined through a J3-P2-3 Junction, only this time the J3 folds were narrower (about 1cm instead of 3cm). The same process of experiment was repeated as in Test 1 with narrower folds. The results were as conclusive as with the first test.

Test 5. More sheets of paper were joined through Peysson Junctions to form an even longer strip of paper, leading to results as good as the first test (see upper image of fig. 4).

Further experiments were conducted varying in paper quality and size, using again a J3-P2-3 Peysson Junction, as in test 1. Variation did not lead to significantly different results. Of course, the use of softer paper (such as paper towels) resulted in the junction coming apart even when little force was employed. A certain degree of paper density seems necessary to ensure junction stays intact. There appears to be a bifurcation threshold, depending on paper density, across which behavior changes drastically. Below the threshold, the two (or more) units of paper come undone by sliding off each other. Above the threshold, the break happens as a result of a tear in one paper unit, away from the junction.

3 Unfolding mathematics with a qualitative twist: Thomism

The tearing of the paper in the above experiments points towards a possible catastrophe, that is, a sudden change in the continuous system. Instead of carrying out a quantitative analysis of paper resistance to stress, we wish to present the reader with qualitative mathematical ideas, and discuss the relevance of folding to mathematics in relation to the paradigms of continuity and discreetness.

The French mathematician René Thom (1928–2002) is associated with two notions of folds and continuity noted above. In 1959 Thom was awarded the Fields Medal for his work in the field of topology[6] (roughly speaking, the study of shapes up to continuous deformations). He later became a major contributor to the related field of singularity theory – the classification of local behavior of various geometrical objects and mappings. He is often associated with *catastrophe theory*, which had its moment in the limelight in the 1970s (more on this in the sequel). Catastrophe theory consists in using ideas from singularity theory to describe the sudden changes that may occur in the qualitative behavior of concrete systems that arise in fields such as biology or economics.[7] It is possible Thom's original ideas were over-interpreted by his followers and epigones.[8] Catastrophe theory was initially not seriously considered; it became a subject of ridicule when it was rather naively adopted by some in order to introduce a little bit of mathematics into the social sciences, as if such proof of rationality was needed to begin with. In fact, catastrophe theory provides a useful conceptual system to analyze phenomena, which were previously understood using more technical, quantitative models. It provides a simple language with which one can interpret and unify disparate phenomena; indeed, Thom was not afraid of simplicity. For him, non-trivial concepts were all the more interesting when they could be expressed in a simple, non-technical form.[9] Later on during the 1980s, catastrophe theory was eclipsed by *chaos theory*, not in opposition to catastrophe theory, but partly influenced by it. Thom was something of a free electron in the French mathematical landscape, at a time when the Bourbaki group exerted the most influence. Bourbaki's radically formalist line ran opposite to Thom's ideas.[10] We shall return to this issue in Section 4.2.

---

6   He was awarded for his work on cobordism theory.

7   A full account of catastrophe theory can be found in Arnold 1992. Let us briefly cite from Tsatsanis 2012, 217: "[catastrophe theory] has as its goal to classify systems according to their behavior under perturbation. When a natural system is described by a function of state variables, then the perturbations are represented by control parameters on which the function depends. This is how a smooth family of functions arises in the study of natural phenomena, [...]" one of its members being a function with critical points.

8   See for example Zeeman 1977 and Smale 1978 for an account of the polemics induced by the popularization of catastrophe theory in the 1970s.

9   Thom signed a disapproving academic report on Jacques Derrida, claiming that his convoluted, difficult language does not hide any deep or subtle thought. See Smith 1992.

10   See for example Thom 1970.

3.1 Ontological anteriority

Of prime importance to us is the *anteriority principle* Thom promoted, namely, the principle of the ontological precedence of the continuum over the discrete.[11] For Thom, discrete objects arise as accidents in the continuum,[12] or they may come about as topological invariants of continuous objects.[13] The anteriority principle reverses the traditional order in which mathematical objects are typically introduced: in the classroom, we are first taught the system of natural and negative numbers $(0, \pm1, \pm2, \pm3, ...)$. The system of rational numbers is then introduced,[14] which is then completed to form the system of real numbers or the real line – the basic continuous object, used, for example, to model our intuition of time. Thom proposed to reformulate mathematics from this viewpoint, starting from what to him was the most fundamental object – the line drawn on a chalkboard.[15] Even though he himself admitted his program could not be carried out formally,[16] he did not see this as a failure, as he often expressed a critical view of the formalization of mathematics.[17]

3.2 Unfolding

How does this reformulation, that is, the anteriority principle, as what commences from the continuous, manifest itself in Thom's theory? The concept of *versal unfolding* in singularity theory that we are roughly going to sketch now is partly based on Thom's ideas; it illustrates, as we shall see, the anteriority principle discussed above. Singularity theory attempts to understand a possibly complicated singular (as opposed to regular) object by embedding it (or unfolding it) in a continuous family of objects of a similar, albeit simpler, nature.[18] By doing so in a suitable manner, one often observes a natural stratification of the parameter space of this family, which reflects the complication of the object corresponding to a specific value of the parameter. The stratification often exhibits a rich structure, which may give some important information on the original singular object under study. An important idea in singularity theory is that one may identify two apparently different objects if one can pass from one to the other using a transformation, which preserves attributes of the objects in question. In many cases, modulo those identifications, one can find an unfolding, which is at the same time not

---

11 Thom 1992.

12 Such as an arbitrarily marked point on a line, or the moment when the long hand of a clock lands at '12'.

13 Such as the number of turns of a rope around a mooring bollard.

14 By 'rational number' we mean a positive or negative fraction such as $3/4$ or $5/3$.

15 Thom 1990. See also Thom 1992.

16 As remarked in Thom 1992.

17 Thom actually promoted the idea that anything stated rigorously is insignificant. Cf. Thom 1968. See also Salankis 1999, sections 4.3 and 4.4.

18 See Thom 1975, especially chapters 3, 4 and 5.

too complicated but general enough to cover all possible qualitative configurations. This is a versal unfolding.[19]

Fig. 5: Fold and cusp.

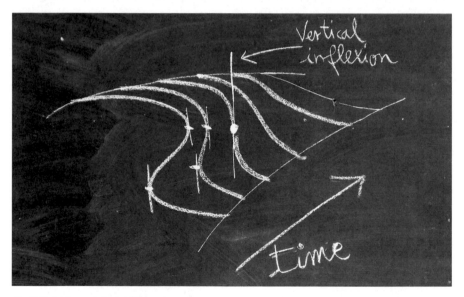

Fig. 6: Unfolding an inflexion.[20]

---

19 Here we understand 'versal' as in 'universal' or 'transversal'. Arnold 1992 is a good introduction to the mathematical definitions of these concepts.
20 The chalkboard illustrations are a homage to Thom, who praised chalk and blackboard.

Emmanuel Ferrand, Dominique Peysson

3.3 Folds and cusps

We now come to a smooth model of a fold ubiquitous in catastrophe theory, which will provide at the same time an illustration of the concept of unfolding. In fig. 5 one sees a blackboard drawing of a local model of a regular mapping from a domain of the plane back onto the plane: a local model of *smooth folding*.

The mapping is defined as follows: the points of the domain are first embedded in three-dimensional space following the picture, and then mapped back to the plane using the vertical projection. Some points in the plane below are the image of exactly one point, while others, as a direct result of folding, are the image of three different points of the embedded domain above. This dual characterization of points defines two regions in the image below – the region of single-source points, and the region of triple-source points. Exceptional points lie at the border between these two regions. The exceptional points form a border curve – a cusped border curve[21] – consisting of two branches, which meet at a special point – a singular point – in the image plane. The set of points in the domain that map onto the cusped curve form a smooth curve.[22]

---

21 A *cusp* is a singular point of a curve, which locally, when located at the origin of the (x,y)-plane, its neighborhood looks like that of the curve $\{y^2 = x^3\}$. The fact that a cusp is created during this projection is noticeable since, in general, the projection of a smooth spatial curve does not have any cusps, but is itself a smooth curve in the plane, possibly with some isolated self-intersections, called *nodes* (i.e. a point that locally looks like the intersection point of two lines). See Hartshorne 1997, 310. When one works over the complex numbers and examines projections of complex surfaces (of degree > 2) into the plane, then every border curve generically must have cusps and nodes (the "border curve" in mathematical jargon is called *branch curve*, being the projection of the *ramification curve*. The ramification curve is, one might say, the curve along which the surface is folded with respect to the projection). For a survey on branched coverings and the special relations between the nodes and the cusps of a branch curve, see Friedman/Leyenson 2011.

22 Note that the number of points in the pre-image of a point along the cusped curve equals two. A more formal description of fig. 5 is as follows. It represents a folded surface, embedded in three-dimensional Euclidean space, R^3, and then mapped back into the plane by projecting in the vertical direction. The surface is parameterized by a domain in the plane, for which we will use (t, s) as coordinates (we assume that this domain is a neighborhood of the origin $(0,0)$). We will map this domain in R^3 using the following formula: $(t,s) \rightarrow (x = t, y = s^3 - ts, z = s)$
The image of the domain under this map is the folded surface in the picture. The real fold is actually achieved when this surface is mapped to the plane by the vertical 'forget z' projection, so that the map from the domain to the image plane takes the following form:
$$f: (t,s) \rightarrow (x = t, y = s^3 - ts)$$
Its differential is not of maximal rank 2 if and only if $3s^2 = t$, i.e. on the parabola (note that when this relation between s and t is substituted for the equation of the surface, we obtain the following curve in R^3, parameterized by s: $s \rightarrow (3s^2, -2s^3, s)$, which would be the *smooth* ramification curve). Outside this curve, it is a local diffeomorphism. The curve is mapped by f onto another curve parameterized by s as follows
$$s \rightarrow (x = 3s^2, y = -2s^3).$$
This curve has a singular point at $s = 0$, called *a cusp*. In the image plane, this curve separates the regions where a point has one or three pre-images.

In thermodynamics a similar picture gives a precise account of the transition from metastable states to stable states. Note that this and was well known long before catastrophe theory came about.[23] Many prototypical examples from catastrophe theory can be examined through the metastable-stable thermo-dynamical paradigm. These examples arise in biology, economics or the social sciences, in contexts in which, at least on a qualitative level, a model involving the minimization of some energy can be introduced.

3.4 Unfolding an inflexion

The mapping between two-dimensional planes, presented in fig. 5, introduces a model of cusp singularity. However, the very same picture also corresponds to an unfolding of a vertical inflexion, as can be seen in fig. 6.

The folded surface (the above domain) can be foliated into curved lines that are parameterized by the real line,[24] as if one is looking at a movie, which tells the story of a curve that develops and changes over time. The curve with the vertical inflexion is singular in the sense that it is qualitively isolated within this family: when it is perturbed the inflexion disappears. Perturbed in one direction, two distinct bumps appear, perturbed in the other direction, there are none. Either way, the perturbed curves run perpendicularly to the vertical direction. The vertical inflection corresponds to the singular moment, where two bumps meet and annihilate each other. Here lies the idea that underlies unfolding: a singular object often corresponds to a limit process, as a boundary or a wall crossing – an accident in continuity. To be understood as such, a singular object cannot be studied on its own, one has to include it in a family of neighboring objects so it could be deformed. From this viewpoint, singularity – and hence also a discrete object, viewed as singular – becomes a relative concept.[25]

---

23   See for example geometrical methods in thermodynamics introduced by Gibbs 150 years ago.

24   The same piece of surface embedded in R^3 can be used to understand the deformation of a vertical inflexion. Using the notation of footnote 22, for any fixed value of the parameter t, which we should now consider as representing time, we can cut a t time-slice of the surface to get a curve parameterized by s in the (y, z) plane. For $t = 0$, this curve $s \rightarrow (y = s^3, z = s)$ has a vertical inflexion: the vertical direction (z axis) is tangent to this curve, but the curve goes through this tangent at the origin. As t varies, the curve is perturbed. For t positive, the curve gets two bumps: the inflexion is resolved into two generic (quadratic) tangencies (i.e. one obtains a function with a minimum and a maximum, to which two lines, parallel to the z axis, are tangent). When t is negative, the curve is transversal to the vertical direction and has no tangency with any line parallel to the z axis. The curve at $t = 0$ is singular in the sense that it is the only one of its kind in this family, i.e. it has only one tangency point with the z axis and it is of multiplicity 3. Perturbing it in both directions gives rise to a curve with simpler tangencies, if any. This family of curves is called an unfolding. The union of all those curves for all possible values of t forms our original folded surface.

3.5 Smooth folding versus paper folding

To make the theory precise, these considerations should be developed in the context of smooth objects.[26] Singularity theories can be applied to describe the physics underlying the experiment leading to catastrophe – the disruption in continuity corresponding to the tearing of our paper strip system. The folds (or curves) involved would then be rather abstract: they would lie within energy levels in some complicated high-dimensional phase space. Nevertheless, one should not confuse this approach, where the idea of folding is abstractly hidden, with our simple paper strip system – a concrete folded sheet of paper.

Fig. 7: Flat folding.

In reference to smooth folding as in fig. 5, it is important to note, that if we were to try to realize a similar cusp singularity in an actual model, we would need for that purpose a plane domain made out of an elastic, stretchable material. A sheet of paper is not a good candidate, as it exhibits little plasticity in this regard. A paper model analogue of cusp singularity will hence look slightly differently. It can be found in the simple singularities that occur in the planar folding of a sheet of paper (as when one folds a city map, see fig. 7). The fold lines of a smooth model are replaced by line segments, along which the sheet of paper is creased. Cusp singularities would correspond to the points where

---

25 We would like to draw attention to Gilles Châtelet (1944–99), a highly original philosopher and pamphleteer, who in *L'enchantement du virtuel,* Châtelet 2010, presented an interesting analysis of an idea closely related to that of unfolding. In the context of his work, 'virtual' should be understood similarly to the principle of virtual work in mechanics, where the static equilibrium of a system (for instance a physical structure such as a bridge) is embedded in some space of possible virtual deformations of its structure. Also cf. Salanskis 2012 for a discussion on the epistemology of Thom's catastrophe theory.

26 Using technical terms, smooth objects are continuous and 'sufficiently differentiable' in a sense appropriate to the given mathematical context.

the line segments meet in the simplest possible way, namely, instances where four line segments meet. One can prove that non-plasticity of paper induces a constraint around each cusp singularity: opposite angles must add up to a flat angle. More complicated non-generic singularities, where a larger (even) number of segments meet, may occur, but one can check experimentally that they do not occur in random flat paper folding.

Nevertheless, complex non-generic singularities are of interest when one designs an origami folding. It would be interesting to further develop its analogy to the classical smooth theory of singularities, in order to understand, for example, how these more complex singularities can be considered a superposition of simpler ones, and how a small perturbation may induce a resolution of the complex configuration into a collection of simpler configurations. To the best of our knowledge, such program has not been carried out yet.

In fact, the smooth setting, which at first glimpse appears more complicated and abstract than simple paper folding, offers many advantages and is quite flexible, once notions of differential geometry are introduced. However, a likely paper folding analogue promises to have more of a discrete flavor: even though a sheet of paper is a continuous media, the resulting pattern of creases is a graph whose edges are segments of straight lines, which is completely determined by discrete set of data.[27]

4 Convoluted literature

In the former two sections, the fold presented the possibility of bridging the distinction and hierarchy between discrete and continuous. It submits itself promptly as a philosophical, as will be discussed shortly in section 4.3. This is also the case with contemporary art as in Roy Ascott's *La Plissure du texte*, with contemporary music as in Pierre Boulez's *Pli selon pli, portrait de Mallarmé*, or with literature as in Joyce's *Finnegans Wake*. In this section "twisted [and] intricately folded" will also mean *convoluted*. Accordingly, in *Finnegans Wake*, we find convoluted characters in folds. As we unfold what the convolution has intricately folded, we reach a higher viewpoint. It is no longer considered speculative nowadays to suggest that creative thinking differs profoundly from Aristotelian logic or a Boolean hypothetico-deductive process. Syncretic thinking, as clearly explained by Anton Ehrenzweig in *The Hidden Order of Art*,[28] is an intuitive and unconscious critical process,

---

27 In the smooth setting, the curves along which the surface is folded (and especially their image when projecting into the plane, i.e. the branch curves) gives rise to combinatorial data regarding the surface itself so that one can reconstruct the covering using that data together with some additional geometrical data. This is the subject of Chisini's conjecture, proved in Kulikov 1999; Kulikov 2008. See also Friedman/Leyenson 2011.

28 Ehrenzweig 1967.

which can give us access to a more powerful vision of complexity. Thinking in terms of convolution offers a modality for this function of thinking. Scientific creativity may come under it, as it requires researchers to think differently as they pretend to when writing their formal articles. We therefore consider how this convoluted viewpoint could provide a different perspective on other fields of the written domain, such as philosophy or poetry.

4.1 Finnegans Wake

In his essay *Ambiviolences: Notes for reading Joyce*[29], Stephen Heath observed that, in interpreting Joyce's *Finnegans Wake*, critics took to two opposite routes. According to one line of interpretation, the text was regarded as absurd and unreadable, while according to the other it was deemed an enigmatic message to be decrypted.[30] Heath himself rejected both approaches: the first as it *a priori* denied the prospect of opening itself up to many fragments of meaning, the second as it missed the creative reading the text calls for. The second approach, which supposes a hidden message, assumes the text to have an immanent unity and continuity – a reading, which Heath believes to be too restrictive. It is rather that the text offers multiple fragments of meaning, which are traced and retraced, colliding and breaking ceaselessly in a textual play that resists homogenization.[31] There is no single style; instead, there is dissolution into a network of modes of expression. Narrative structure is discontinuous; narration is ambivalent, discouraging an attempt to identify a specific voice. Heath uses the analogy of the fold to describe this peculiar construction: "[...] Finnegans Wake is offered as a permanent *inter*plication, a work of folding and unfolding in which every element becomes always the fold of another in a series that knows no point of rest."[32] *Finnegans Wake* should not be read as assuming a permanent state of meaning, and reading becomes a creative experience for the reader. The novel is an open piece of art. Referring to Umberto Eco, this means it cannot be reduced to a single interpretation.[33] The text is an endless source of meanings. Beyond a purely informative discourse, the folded and refolded text – convoluted – stimulates the expectations of the reader.

---

29 Heath 1984, 31 – 68.
30 Ibid, 31.
31 Ibid, 31, 32.
32 Ibid, 32 (italics is in the original).
33 Eco 1965, 20.

4.2 Crumpling scientific papers

Against this background of a serial unfolding "that knows no point of rest" and resists any clear, distinct meaning, thereby affording unfolding a flux of meanings, consider scientific papers, which present themselves as examples of an informative text designed to have a clear, linear, discrete reading, without creases or folds, in short, unambiguous (an assertion which itself would be interesting to question). However, creativity in science cannot be reduced to this type of process. The mathematician Henri Poincaré tried to explain in *La Valeur de la science* how much his own discoveries owe to the wandering of thought and daydreaming.[34] Due to present-day academic norms,[35] the production of hard-to-digest, formal, technical literature is experiencing exponential growth. Computer-generated mock-up scientific texts are accepted for publication in scientific journals and conference proceedings.[36] Scientific knowledge is not reducible to the content of scientific texts. Should all scientists disappear from this planet, it is very unlikely science could be reconstructed from archived scientific texts. One could say Joyce's writing is parallel to true scientific practice, in that it gives us a way to perceive the indistinct, the random, the discontinuous in seemingly linear flows.

The Bourbaki project is a singular milestone in the landscape of formal mathematics and scientific writing.[37] Bourbaki was a collective of mathematicians who undertook to provide modern mathematics with a formally sound foundation through a unified approach.[38] Although a great deal of work was done, leading to many useful results, the project remains a fascinating but unattainable ambition. Eighty years since, one is bound to admit this endeavor is beyond human capabilities, as few can absorb mathematics this way. Thom, who was mathematically active as the Bourbaki project reached its zenith, steered away from this circle of ideas. Interestingly, differential calculus, upon which Thom's smooth paradigm is based – a paradigm invoked when one wishes to establish a formal account of the continuum and of singularity theory – was developed by two convoluted writers: Isaac Newton (whose abundant obscure writings extend far beyond science) and Gottfried Wilhelm Leibniz.[39]

---

34   Henri Poincaré wrote a whole paper about this process. See Poincaré 1908.
35   The so-called publish-or-perish imperative.
36   Several papers randomly generated by Mathgen were accepted for publication after a peer review process. See *http://thatsmathematics.com/mathgen*
37   Bourbaki 1939.
38   A good introduction to Bourbaki can be found in various works by the French poet Jacques Roubaud. See Roubaud 1997.
39   Leibniz himself uses the fold as a metaphor to a more or less explicit degree: Deleuze famously noted in reference to Leibniz's monadology: "In the labyrinth of the continuous the smallest element is not the point but the fold." See Deleuze 1988, 9.

4.3 Folding mystery

Many authors utilized the fold as a metaphor: Martin Heidegger (the fold of our being), Maurice Merleau-Ponty (the fold as chiasmus or interlacing), Michel Foucault (the fold of in and out), Gilles Deleuze (surface fold), Jacques Derrida (especially in his writing about Joyce, the fold as the impossibility of simple self-identity) and Jean-Luc Marion (fold of the given), to name a few.[40] This goes beyond the scope of this paper; here we will merely mention the work of Quentin Meillassoux[41] on Stéphane Mallarmé's seminal poem *Un Coup de Dés Jamais N'Abolira Le Hasard*.[42] This poem is a landmark of French literature, possibly of the entire literary corpus. Published at the very end of 19th century, it represented a radical rupture: the poem, whose form in print is of primary importance, exhibited an intriguing interplay between the two-dimensional page and its non-linear typographical layout. Its syntactical structure was highly non-standard, in particular in its use (or lack of use) of punctuation. With about 700 words per 11 pages, a vast majority of the paper surface was left blank. The words, printed using various font sizes, were laid out in a succession of clusters, which may evoke foam on the crest of waves. Meaning, if indeed there is any, has remained mysterious for more than a century; the poem has been the subject of countless studies and interpretations. Like in the case of Joyce's *Finnegans Wake*, it has been considered absurd and unreadable by many, or, alternately, a challenging encoded mystery by others. Quentin Meillassoux recently proposed a cryptologic interpretation,[43] largely based on a certain 'discrete' numerology. More precisely, it involves a careful consideration awarded to all words and signs in their respective positions – the most basic 'discrete' elements of text. Meillassoux's precise, quantitative, documented work ignores in effect most of the poem's 'geometry' (layout, varying sizes, global and local blank regions). Meillassoux is nevertheless able to draw rather convincing conclusions from his analysis. Roughly speaking, some numbers are not there by accident; it seems that Mallarmé adopted a hidden structure governed by obscure computations. We note that this discovery, if confirmed, is but one interpretation that, albeit appealing, still disregards more material, visual aspects.[44] We would like to suggest a more 'folded' approach, one possibly based on manipulations (yet to be discovered) of the poem's underlying sheet of paper used here as the two-dimensional

---

40　For these fold metaphors see Cormann/Laoureux/Piéron 2005.

41　Meillassoux 2011.

42　Mallarmé 1897.

43　Kaplan 2012.

44　Poetry was for Mallarmé a form of a new religion, encompassing other forms of artistic expression: it has its own musicality, and also in this poem, a material and a visual representation. See Meillassoux 2015, 65, 104.

space of poetry.[45] Even if an actual act of folding is abandoned, we feel that the inherent folded nature of those clusters of words – suggesting waves of a rough sea – should be taken into account. It seems to us that a third folded approach should be followed similar to the one suggested by Heath: the words themselves would be folded and unfolded onto each other. This would not substitute the intriguing search for the ultimate (discrete) decryption à la Meillassoux, but would run parallel to it.[46]

Looking at the history of ideas, one observes that the fold, as a metaphor, has often been used in philosophical dialectical contexts. It has been invoked to deal with single and multiple, identity and difference, inside and outside or the constituent and the constituted. The image of the fold appears to be an effective metaphor whenever one contemplates complexity. As evidence, one only need consider its many occurrences in contemporary philosophy. Creative thinking is based on a complex relationship between all manner of rational and irrational thought. In that sense, convoluted images support creative imagination. The writings of Joyce, Thom and Mallarmé invite a creative reading that allows the reader to follow convoluted paths and nebulous themes.

Our existing models of complexity required revision, which prompted the emergence of chaos theory and of the notion of a complex system. Folds, convolutions, windings and labyrinths are all fruitful methodologies that help to structure, represent and rethink the complexity of our world. We followed here some folded dualities of our own, in a discussion starting from a folded paper junction. The fold has often been associated with infinitely divisible continuity. Our physical performance has taken an opposite direction, creating a structural junction from the discrete to the continuous. Our bodies were fully engaged in a dynamic one-dimensional experiment, experiencing the moment when the continuous broke into the discontinuous. The sudden toppling over of one of the performers brought on catastrophe. In the same vein, we presented a patchwork of mental images, dealing with seemingly unrelated subject matters, hoping readers will find their way to continuously combine them.

---

45   Mallarmé gave much thought to the layout of book leaves in the folio. Different combinations of paper-sheets suggested rhymes. The book binding process was important to Mallarmé, who considered the fold to be the central abyss, the "discontinuity of meaning," the nonsensical part associated to text signification. The white part of the page he thought of as the void. Se e Meillassoux 2015, 59.

46   Randomness (*le hasard*) was of great significance to Mallarmé. See Meillassoux 2015, 39.

5 Appendix: instruction for a printed performance

The sense of mystery suggested by the labyrinthine convolutions of folds is crucial to their nature. The fold can hide what has been laid down on paper. It offers an effective way to encode messages. We explore this theme here in a proposed folding experiment. We encourage the reader to print and fold the very last figure (fig. 8) following our instructions below. It brings us back full circle to Henri Michaux, the poet and painter cited at the very beginning of this paper, as the experiment refers to a modified (unnamed) ink print of his – this paper's last figure.[47]

Note the markings at the top and bottom of the very last image: points, thin long segments, thick short segments. The reader will need to carefully fold the sheet of paper several times parallel to the long edge. The paper should be folded in 'V' shape (a valley fold) across each line that joins a matching pair of thin long segments – a line that runs parallel to the long-edge of the paper. The resulting fold should stretch across the entire length of the paper from top to bottom. In a similar manner, the paper should be folded in '∧' shape (a mountain fold) across each line that joins a matching pair of thick short segments. Once this is done, each thick line should touch a corresponding point – the continuity point. The folds should be carefully flattened. Once all folds are accomplished, the reader will observe the result across the flattened width of the paper.[48]

---

47 Henri Michaux, Sans titre, ca. 1975, ink on paper (mongrammed at lower right), 74,5 x 105 cm, Belgium, Private Collection. This piece by Henri Michaux will be included in the future in a catalog prepared by Micheline Phankim, Rainer M. Mason and Franck Leibovici. The catalog will present prints of his paintings and drawings, the majority of which feature in private collections.
48 Should the reader find the printed image too small to fold, he or she may download the original at: *http://webusers.imj-prg.fr/~emmanuel.ferrand/fold/*.

Emmanuel Ferrand, Dominique Peysson

Bibliography

Arnold, Vladimir Igorevich (1992): **Catastrophe Theory.** Berlin: Springer Verlag.

Bourbaki (1939): **Eléments de mathématiques.** Paris: Hermann.

Châtelet, Gilles (2010): **L'enchantement du virtuel.** Paris: Rue d'Ulm.

Cormann, Grégory / Laoureux, Sébastien / Piéron, Julien (eds.) (2005): **Différence et identité. Les enjeux phénoménologiques du pli.** Paris: Vrin.

Deleuze, Gilles (1992): **The Fold. Leibniz and the Baroque.** Trans. by Conley, Tom. Minneapolis: University of Minnesota Press.

Ehrenzweig, Anton (1967): **The Hidden Order of Art. A Study in the Psychology of Artistic Imagination.** Berkeley: University of California Press.

Eco, Umberto (1965): **L'œuvre ouverte.** Paris: Sueil.

Ferrand, Emmanuel / Peysson, Dominique (2015): **Onto the fold.** Performance given as part of the workshop *Folding and Creasing of Thin Plate Structures* Ecole supérieure de physique et de chimie industrielle (ESPCI), Paris, March 26.

Friedman, Michael / Leyenson, Maxim (2011): **On ramified covers of the projective plane I: Segre's theory and classification in small degrees (with an appendix by Eugenii Shsustin).** In: International Journal of Mathematics, vol. 22, no. 5, pp. 619 – 653.

Hartshorne, Robin (1999): **Algebraic Geometry.** New York: Springer.

Heath, Stephen (1984): **Ambiviolences: Notes for Reading Joyce.** In: Attridge, Derek / Ferrer, Daniel (eds.): Post-structuralist Joyce. Essays from the French. Cambridge: Cambridge University Press, pp. 31 – 68.

Kaplan, Edward K. (2012): Review of Meillassoux, Quentin (2011): **Le Nombre et la sirène. Un déchiffrage du «Coup de dés» de Mallarmé.** Paris: Fayard. In: Nineteenth-Century French Studies, vol. 40, no. 3, pp. 343 – 344.

Kulikov, Viktor (1999): **On Chisini's Conjecture.** In: Izvestiya: Mathematics, vol. 63, no. 6, pp. 1139 – 1170.

Kulikov, Viktor (2008): **On Chisini's Conjecture II.** In: Izvestiya: Mathematics, vol. 72, no. 5, pp. 901 – 914.

Mallarmé, Stéphane (1897): **Un coup de dés jamais n'abolira le hasard.** Paris: Cosmopolis.

Mathgen (website): Online: http://thatsmathematics. com/mathgen (last access: December 8, 2015).

Meillassoux, Quentin (2011): **Le Nombre et la sirène. Un déchiffrage du «Coup de dés» de Mallarmé.** Paris: Fayard.

Michaux, Henri (1941): **Au Pays de la magie.** Paris: Gallimard.

Poincaré, Henri (1905): **La Valeur de la science.** Paris: Flamarion.

Poincaré, Henri (1908): **L'invention mathématique.** In: L'Enseignement Mathématique, vol. 10, pp. 357 – 371.

Roubaud, Jacques (1997): **Mathématiques:.** Paris: Le Seuil.

Salanskis, Jean-Michel (1999): **Le constructivisme non standard.** Villeneuve a'Ascq (Nord): Presses Universitaires du Septentrion.

Salanskis, Jean-Michel (2012): **Métaphysique et épistémologie de la catastrophe.** In: Critique, vol. 8, no. 783 – 784, pp. 687 – 698.

Smale, Steven (1978): Review of Zeeman, Erik C. (1977): **Catastrophe Theory, selected papers 1972 – 1977 by E. C. Zeeman.** London: Addison-Wesley. In: Bulletin of the AMS, vol. 84, no. 6, pp. 1360 – 1368.

Smith, Barry et al. (1992): **Letter to the Times.** In: The Times, 9 May.

Thom, René (1968): **La science malgré tout.** In: Encyclopaedia Universalis (Organum), vol. 17, Paris: Encyclopaedia universalis France, pp. 5 – 10.

Thom, René (1970): **les mathématiques "modernes": une erreur pédagogique et philosophique?.** In: L'âge de la science, vol. III, no. 3, pp. 225 – 236.

Thom, René (1975): *Structural Stability and Morphogenesis: An Outline of a General Theory of Models*. Trans. by Fowler, David H.. Reading et al.: W. A. Benjamin.

Thom, René (1990): *Un modèle continu du nombre*. In: Les rencontres physiciens-mathématiciens de Strasbourg – RCP25, vol. 41, pp. 163–189.

Thom, René (1992): *L'Antériorité Ontologique du Continu sur le Discret*. In: Salanskis, Jean-Michel / Sincaeur, Hourya (eds.): Le Labyrinthe du Continu. Berlin: Springer, pp. 137–143.

Thom, René (1993): *Prédire n'est pas expliquer*. Paris: Flamarion.

Tsatsanis, Peter (2012): *On René Thom's significance for mathematics and philosophy*. In: Scripta Philosophiae Naturalis, vol. 2, pp. 213–229.

Zeeman, Erik Christopher (1977): *Catastrophe Theory, selected papers 1972–1977*. London: Addison-Wesley.

Emmanuel Ferrand
Email: *emmanuel.ferrand@upmc.fr*

Institut de Mathématiques de Jussieu-Paris Rive Gauche, Université Pierre et Marie Curie (Paris VI) France, 4 place Jussieu, 75005 Paris, France.

Dominique Peysson
Email: *dominique.peysson@cegetel.net*

École Nationale Supérieure des Arts décoratifs (ENSAD)
31 rue d'Ulm, 75240 Paris, France.

Sandra Schramke

3D Code: Folding in the Architecture of Peter Eisenman

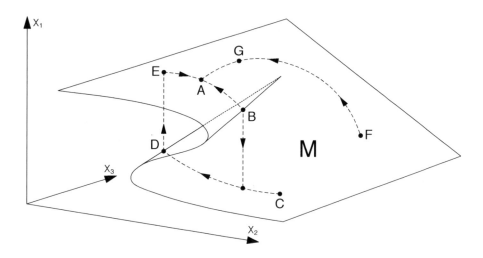

Fig. 1 Chaos theory, René Thom.

"A 'cryptographer' is needed, someone who can at once account for nature and decipher the soul, who can peer into the crannies of matter and read into the folds of the soul."[1]

In the 1990s Peter Eisenman (b. 1932), who since the 1960s has been regarded as one of the most advanced contemporary theorists and practitioners of architecture, developed with the form of the fold a new architectural language based on texts as well as on mathematical models. He supported this with Gilles Deleuze's (1925–1995) book *Le Pli* (1988), new phenomenology – as presented for instance by Gernot Böhme (b. 1937) in *Atmosphäre: Essays zur neuen Ästhetik* from 1995 – and René Thom's (1923–2002) chaos

---

1   Deleuze 1993, 3.

theory.[2] For Eisenman the fold is a form that bridges inner and outer worlds, geometry and the description of mental states. At the same time, the fold is able to build on the geometry of the grid as well as on the theories of perception and mathematics. Historically, the fold can be traced back in particular to the metaphysics of Leibniz's *Monadology*.[3] With the example of the drapery characteristic of the Baroque age, Leibniz (1646–1716) searched beyond systematic rationality for new concepts of singular phenomena, not in the logic of an object, but in accordance with the idea of an "aesthetic, [...] poetic thing."[4] In his *Monadology*, which has come to be known as a theodicy, Leibniz thus distanced himself from contemporary philosophical currents of atomistic and substance theories.[5] Rather, he translated the phenomenon of folded cloth into a general concept and linked this to the idea of the universe, which he "describes as a continuous body that is 'not divided, but transfigured in the manner of wax, and folded in various ways like a tunic.'"[6] In this way, Baroque drapery as formalized matter was able to avoid the accusation of the mere decoration of the clothed body and, with reference to Aristotle's theory of elements, be transformed into a mental object. As Deleuze notes, Leibniz's metaphysics is based on a conception of the world that is oriented not to Descartes's (1596–1650) unity of the manifold, but to a multiple of the event. Here, the event is materialized in the form of folds, which according to Deleuze are actualized in the soul. In the following, the influence of the theory of monads on Eisenman's architecture will be explored. The argumentation follows on the one hand the logic of the geometry of the fold, which as a Euclidean projection displaces the form of the grid, and on the other the logic of American pragmatism, whose history has given rise to a variety of forms, including, in particular, tensegrity structures. These forms in turn can be interpreted as direct precursors to the fold. To finish, I will examine the theme of the fold's space of representation, which in Eisenman's architecture not only takes account of standards of communication, but also brings about a singular affective space – one that is characteristic of the style of the architectural fold.

Geometry of the Fold

Eisenman thus takes up the neoplatonic tradition of the theory of monads and transfers the fold onto his architecture. He situates this form, which is so characteristic of his designs of the 1990s, between body and soul, geometry and perception. At the same time, the fold in its genesis draws on the grid as one of the key foundations of architectural design. The fold makes the grid a condition by undertaking on the basis of the

---

2   Böhme 2013, 385–412.
3   Strickland 2014.
4   Vogl 2003.
5   Adams 1998.
6   Bredekamp 2004, 16.

gridded surface irregular, singular displacements of the same. The grid thus becomes the elementary point of reference from which Eisenman simultaneously detaches the fold. These irregular, singular displacements within the grid are articulated by Eisenman with the idea of an architecture of the *informe*.[7] Furthermore, Eisenman understands the fold as a connection to the outside resulting from the removal of the distinction between figure and ground. In this connection, Eisenman makes use of language, to in turn with reference to Deleuze describe ex-pli-cation as both concept and action in the process of folding and unfolding. To give the form its own ontological status while simultaneously bypassing the logic of the grid, Eisenman replaces the concept of the grid with that of the frame, which he borrows from a book by his long-term friend Jacques Derrida (1930 – 2004), *The Truth in Painting*.[8] With the frame Derrida examines with the help of the garment and with reference to Immanuel Kant the dichotomy between *ergon* and *parergon* in order to explore the difference between an object's in- and extrinsic qualities. Along the division between inside and outside, the frame in this assessment can be perceived as a marking with varying degrees of formalization. Derrida in *The Truth in Painting* gives the example of columns around a building, with which "is announced the whole problematic of inscription in a milieu, of the marking out of the work in a field of which it is always difficult to decide if it is natural or artificial and, in this latter case, if it is *parergon* or *ergon*."[9] With reference to Derrida, Eisenman develops the image of a subordinate dimension of the folding of the space in which this folding is inserted. According to the philosopher and architectural theorist John Rajchman (b. 1946), this insertion in space can itself be considered as active geometrization, but which in distinction to origami is not reduced to a figure, but incorporates the milieu as part of the folding. This is realized by Eisenman for example in his design for Rebstockpark, in which he crosses over the fixed limits of the site to include what is external to the site in the folding: "The Rebstock fold is thus not only a figural fold as in origami – not a matter simply of folded figures within a free container or frame. Rather the container itself has been folded together, or complicated, with the figures."[10]

Here, the central question circles around the potential of the peripheral, which has also been touched on by Aby Warburg (1866 – 1929) with his concept of the "bewegtes Beiwerk" (animated accessory). In this context, Eisenman pursues with the concept of the index taken from linguistics the idea of a potential available to the fold found in the concepts of ex-pli-cation and un-folding. In this way, architecture, understood as text, can be understood as the constitution of the absent.[11] Here, Eisenman pursues a conceptual thinking with regard to the diagram, which on the basis of the different

---

7   Eisenman 2007b.

8   Derrida 1987.

9   Derrida 1987, 59.

10 Rajchman 1998, 21.

11 Eisenman 2004a, 299f.

possibilities of classification and its interstices, moves along the dividing line between order and disorder, complexity and the reduction of complexity: "The diagram is indeed a chaos, a catastrophe, but it is also a germ of order."[12]

The diagram thus offers another approach to Leibniz's metaphysics. If the latter was still characterized by the idea of divisions, Deleuze reverses the idea of unity into that of the multiple: "Classical reason toppled under the force of divergences, incompossibilities, discords, dissonances. But the Baroque represents the ultimate attempt to reconstitute a classical reason by dividing divergences into as many worlds as possible, and by making from incompossibilities as many possible borders between worlds."[13] With reference to Deleuze's concept of the multiple, Eisenman also clearly disassociates himself from the much-quoted mathematician and architectural theorist Leon Battista Alberti (1404–1472), who in the play with geometries described their functional unfolding.[14] In contrast to Alberti, who derived all variations of the multiple from the ideal and real grid, Deleuze defines the multiple inversely as a condition of unity: "The multiple is not only what has many parts but also what is folded in many ways."[15] Thus, while Alberti's considerations take unity as their point of departure, for Deleuze there is no finite unity; rather, there are endless possibilities of the multiple in the unfolding. Unlike Alberti therefore, Deleuze's considerations on the fold do not result in a universal model. Neither Cartesian space nor Aristotelian place is accepted into his approach.

The Inclined Plane as Precursor to the Fold

Eisenman's architecture of the fold, in its concretion, can, however, be traced back to direct precedents, which were initially still characterized by the geometry of the plane. In his manifesto *The Oblique Function* from 1966, the architect and philosopher Paul Virilio (b. 1932) distances himself from modern architecture insofar as this is based on a separation of functions.[16] Virilio reverses this principle. Together with the architect Claude Parent (b.1923) he developed at a distance to modernism the form of the inclined plane, which unified functions instead of separating them. Virilio and Parent both derived the form of the inclined plane from their perceptual experiences in bunkers during World War II, which they combined with the concept of a multiplication of life taken from developmental biology, and applied to the spatial layout of their design for Maison Drusch from 1963 near Versailles.[17] As with a stage, the inclined plane provides a possible space for actions.

---

12   Deleuze 2003, 102.
13   Deleuze 1993, 81.
14   Alberti 1454, 16f.
15   Deleuze 1993, 3.
16   Virilio 1997.
17   Virilio 1997. Cf. also Maak 2010, 20.

Hence, at the center of this architecture are not fixed allocations or occupations of place and space following a Platonic model, but events linked to time. In comparison, Deleuze alludes in the concept of the event – which he derives from Leibniz's theodicy,[18] and originally described the demise of God's omnipotence in the modern period – to the loss of hierarchically structured centeredness.[19]

The event is defined with regard to the fold as a phenomenon of decentralism, which in geometrical terms represents a deviation from the regularity of the grid. To understand this better, it is worth looking briefly at the origins of the grid.

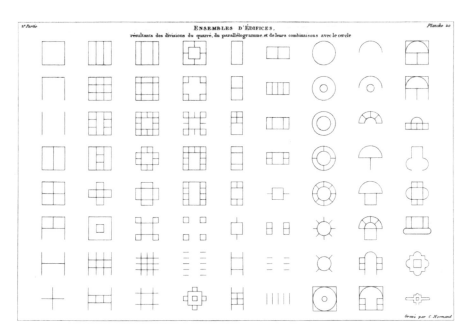

Fig. 2 Ensembles d'édifices, Jean-Nicolas-Louis Durand.

The Origin of the Grid

In 1803 at the École polytechnique in Paris, the French revolutionary architect Jean-Nicolas-Louis Durand (1760–1834) introduced the schematism of a plan grid in the form of graph paper, which provided the basis for systematic geometrization. Durand used the grid to develop model books for elementary types of functional buildings.[20] True to militant enlightenment ideas, the clear and evident elements were intended to establish a lasting

---

18   Leibniz 1978, vols. I–VII.
19   Eisenman 2004b, 235f.
20   Durand 2000.

architectural language. Indeed, Durand anticipated important parts of the functionalism of the beginning of the 20th century, as well as Ernst Neufert's *Bauentwurfslehre*. However, not all recipients were pleased with such practices. Gottfried Semper (1803–1879) for example, as a representative of a tendency opposed to neoclassicism, criticized the schematism in Durand's square grid, in which he feared the end of architecture. Durand operated in the spirit of 19th century scientism, in which earlier concepts of experience-based subjective judgment would be replaced by the modern concept of a universal science. Nevertheless, he was not the first to make use of such ideas.

Already 50 years earlier the teaching of geometry had been introduced at the École du Génie de Mézières, which had given rise to the teaching of stonecutting based on drawing. This was characterized by a representation of stone constructions using parallel projection, and thereby represented an application of stereometry. The resulting *trait*, the term for the production of templates used in cutting the stones, could only be communicated with difficulty, however, and therefore was not able to establish itself as a symbolic language.[21] In the shift in training from general geometry to a drawing to be produced according to fixed rules, Alois Riedler was able to recognize the new possibilities of such a universal language for engineers and architects.[22] Just as already in the 18th century the career of the engineer could establish itself with a focus on mathematics, and subsequently the further specialization of engineer and architect made the profession of master builder obsolete, drawing became the most important medium of communication for architect and engineer. Against this background, graph paper gave rise to a particular form of systematic training in drawing, which has another precursor in the techniques and instruments of surveying.[23]

This can be traced back as far as the Romans, who built their streets and cities on the basis of geometric grids. Starting from the division of the sky, which was then projected onto the ground as *umbilicus* or navel, the Romans determined the *cardo* and *decumanus*, the east–west and north–south axes, thereby forming four quadrants. These quadrants were then repeatedly divided until they assumed the dimensions of the floor of the Pantheon.[24] Besides this form of an idealized division, the quadrants thus obtained also formed the economic units of their calculated administration.

The architectural historian Volker Hoffmann attributes the invention of a net composed of threads to the master builder Filippo Brunelleschi (1377–1446), insofar as he is able to show that the latter measured the topography of ancient Rome with the help of

---

21   Evans 1995, 179–238.
22   Riedler 1896.
23   Grelon 1994, 15–57.
24   Sennett 1994, 106–108.

squared paper.[25] In this way, Brunelleschi affiliates himself in his methodology with the contemporary astronomer Paolo dal Pozzo Toscanelli (1397–1482), who likewise made use of gridded paper to map constellations, and thereby called on the long tradition of employing grids in cartography since antiquity. In his *Trattato di architettura* the sculptor, master builder, and theorist Antonio Averlino (1400–1469), known as Filarete, also points out the advantage of squared paper for the securing of true-to-scale representations.[26]

In the art of the Renaissance the grid was subsequently developed for the purpose of faithful depiction in the graphic arts. Even before Dürer's (1471–1528) classic text, the architectural theorist Leon Battista Alberti as well as the painter and polymath Leonardo Da Vinci (1452–1519) made analogous observations with regard to the *velum*.[27] In 1528 Albrecht Dürer finally used a *velum* as the lattice between the draftsman and his model in his woodcut *Der Zeichner des liegenden Weibes*. Under the influence of Euclid's *Elements*, Dürer formulated in the *Unterweysung der Messung* his methods for a systematic training in geometry.[28]

The Dynamization of Geometry

While Alberti still strove to fix an object in its position in space by means of a *velum*, as was also suggested by René Descartes in his theory of the distinction between *res extensa* and *res cogitans*, in the modern physics of the 20th century at the latest there was a shift in the relation to space and time: time took primacy over the previously fixed metric order of the spatial context.[29] An intermediary step in this direction was taken by Leibniz with his theory of monads, in which he opposed the mechanistic physics of Descartes, in order with the help of point-like centers of force to formulate a unity of non-material, metaphysical conditions. In this connection, Horst Bredekamp has described Leibniz as a "symbol maker par excellence."[30]

If in their beginnings the factors of verifiability and measurability still served an emancipation from the hegemony of theological and aristocratic forms of knowledge and power, then subsequently, as an attempt at universal topological as well as physiological forms of measurement, they represented a new formation of geometrization and disciplining.[31] Thus, the English philosopher and physician Robert Fludd (1574–1637) – two generations

---

25 Hoffmann 1992, 323f.
26 Spencer 1965, 315.
27 Alberti 1435, 16f.
28 Dürer 1538.
29 Koyré 1990.
30 Bredekamp 2004.
31 Galilei 1610. Cf. also Zur Lippe 1974.

after Leonardo da Vinci – at the beginning of the 17th century searched for modern forms of the measurability and representability of physiological coherences between the outer world and the inner human. Here, he moved between fully achieved painterly symbol and graphic technical sketch.[32]

American Pragmatism

The beginnings of a systematic recording of the world were not, however, related only to art and technology, but also to natural phenomena. Around 1900, with his atlases of art forms in nature, in which in the form of symmetries he presented the theory of evolution as a worldview, the zoologist and physician Ernst Haeckel (1834–1919) paved the way for biologism.[33] With the beginning of the 20th century, a new orientation took place with the mathematician and biologist D'Arcy Wentworth Thompson (1860–1948). In his magnum opus *On Growth and Form* Thompson examines the influence of mathematics, physics, and mechanics on evolution against the background of thermodynamics.[34] With his mechanistic explanatory models he assigned, on the basis of mechanics, causal explanatory models to evolutionary morphological changes in form, and thus developed further a Cartesian conception of the world. What concerned him were the ways in which thermodynamics could be transferred to matter, while the immaterial disappeared entirely from possible explanatory approaches. According to Thompson, changes in form comprised both the conception of Cartesian transformations and ideas of morphogenesis.

The Transformation of Tensegrity Structures

These new models of transformation have been featured by architecture in different forms. Perhaps the most prominent was developed by the engineer and architect Richard Buckminster Fuller (1895–1983) at Black Mountain College, founded in 1933 in North Carolina, USA, which made a name for itself through the program of an equal status of art and science in the service of social responsibility. The reform-pedagogical ideas of the philosopher John Dewey (1859–1952), who at his institute at Black Mountain College focused on experience-based interdisciplinarity, contributed decisively to the formation of the school of functionalism. These ideas should be understood as a consequence of American pragmatism, which after the turn of the century viewed itself as a "Reform

---

32   Fludd 1619.
33   Haeckel 1913. Cf. also Breidbach 1998.
34   Thompson 1917.

Darwinism."[35] With the figures Dewey, William James, and George Herbert Mead, pragmatism was able to enter into a union with functionalism.[36]

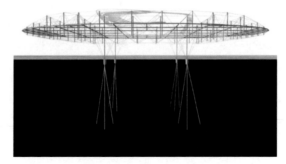

Fig. 3 Blur pavilion, tensegrity structure, computer model, Diller Scofidio + Renfro.

Buckminster Fuller operated in the spirit of an architectural conception mediating between pragmatism and functionalism that is typical especially of the American 20th century. With the example of the Blur pavilion by the American architects Elizabeth Diller and Ricardo Scofidio for the Swiss Expo 2002, it can be shown how the technical basis of the pavilion supports the idea of the fleetingness of the image of a cloud.

Here, the transformation of appearance enters into a union with engineering, which can be traced back to Buckminster Fuller's form experiments. While the pavilion as walk-in cloud takes up traditional technical and allegorical theories of space, and stages them anew, its floor is made of a tensegrity construction made of steel tubing with flexible joints and 20,000 mist nozzles. The architects, who decided to work with the surrounding water, which was sprayed in the form of mist by means of a computer-controlled weather system, entirely abandoned the architectural categories of inside and outside. The idea of a skin is replaced by an atmospheric space of experience. The play with the two states of the fluid and the gaseous involved an outer and optically perceivable space of vision as well as an inner and atmospherically experienceable affective space. Interesting in this construction is the pavilion's undulating floor, which as load-bearing system symbolizes the instability that all transformations of the figure contain in themselves. A tensegrity structure as a suspension system made of flexible tension rods formed the technical basis for the aesthetic appearance of this cloud. Its ingenious inner and outer balance in the tradition of American functionalism after Buckminster Fuller followed the logic of an optimized reduction of supporting elements. As lightweight load-bearing structure

---

35   Goldman 1997.
36   Hutterer 1998, 181.

it could be installed in an uncomplicated way and with a minimal use of materials. On a structural level, stable polyhedra acted as the basic elements for this load-bearing construction.

Buckminster Fuller took as a starting point a dynamic balance of swaying and interconnected systems, and not a grounded structure. With regard to its dynamic balance of forces, this structure is closely related in a figurative sense, on the one hand, to the model of space-time in modern physics, which abandoned fixed physical quantities and translated them into vectors; on the other, to the aesthetic forms of natural orders such as snowflakes and crystals.[37] The genealogy of tensegrity is worth at least sketching out. The term is a neologism composed of the terms tension and integrity. The first tensegrity structure was the wire-spoke wheel developed between 1871 and 1876 by James Starley (1830 – 1881). Thus, when in 1913 Marcel Duchamp mounted a bicycle fork with its wheel on a stool, and presented the result as the first Readymade, he was simultaneously showing an object that was also worth looking at due to the balance of forces contained in it.

The further development of this principle was carried out by Buckminster Fuller and his student, the artist Kenneth Snelson (b. 1927).[38] Tensegrity structures are characterized by a load-bearing system held in balance by a series of elastically connected tension rods. This dynamic structure gives rise to a form that cannot be described with the idea of groundedness found in traditional architecture. This was demanded by the Roman architect Vitruvius in his De architectura from the 1st century B.C. as one of the defining categories on the basis of the Aristotelian conception of the solid Earth. Since then it has dominated the history of architecture in the form of a rectangle. As Ernst Cassirer has noted, this category, which is based on a fixed concept of space, underwent a transformation at the latest during the Renaissance period with the transition from an Aristotelian to a dynamic conception of the world. Immanuel Kant would then detach this relation from space and dynamize it in the direction of time.[39] In the 20th century this change is exemplary of the change in the conception of the physical order brought about by Einstein's theory of relativity, which describes the universe as a field consisting of a balance of forces.[40] During the final years of World War II a number of researchers and engineers began to think over the basic structure of engineering in a model setting. On this basis, Buckminster Fuller became interested in a newly founded, universally valid engineering based on not two- but three-dimensional geometry, in which two lines do not intersect at a point as on paper, but form two linked elements in space with a specific relation of forces. He associated this practical geometry with a particular interpretation

---

37   Fuller 1982.
38   Snelson 2015.
39   Kant 1786.
40   Cassirer 1956.

of the transitions of the elements described by Plato in *Timaeus*, the so-called platonic solids.[41] Among other things, Fuller reconstituted the Platonic solids. He began with a square whose sides were made of rigid struts, but whose corners were made of flexible rubber tubes. This led to astonishing results. Of its own accord the square does not keep its shape, but topples over like a slanting cube. The figure subsequently passes through various states of oscillation; the only stable form here is formed by an isosceles triangle. Hence, it was this rather than the square frame that Fuller designated as the basic element of his building structures.[42]

Through his interest in the geometries of Platonic solids and other original experiments, Buckminster Fuller was alerted to the physically effective structural laws of nature. He subsequently tested his results in empirical experiments on various load-bearing systems. The first of these experiments can be traced back to the early 1940s. However, it was only after World War II that they were given a memorable shape in the practical forms of geodesic cupola and dome, which would eventually make Buckminster Fuller famous.[43] His constructions were based on a flat and convex or concave space frame, the so-called octet procedure made up of alternating octahedron and tetrahedron cells.[44]

With this transferal of scientifically-oriented model questions and the structure of atomic configurations onto load-bearing systems, Buckminster Fuller was turning in particular against the International Style. In his introduction to *Five Architects*, Colin Rowe describes the International Style as a reflection on form in the context of economic agendas especially in the USA.[45] The outer sign of this official direction was a loss of the ideals of European architectural modernism. The related tendency of the reduction of architecture to its form is opposed by the so-called New York Five – to which besides Michael Graves, Charles Gwathmey, John Hejduk, and Richard Meier also Peter Eisenman belonged – with the help of a semiotics derived from linguistics. With the institutionalization of the subject of architectural theory and the application of semiotics to architecture in accordance with the linguistic turn diagnosed by Gustav Bergmann and Richard Rorty, the New York Five searched for new, more open interpretative models.[46]

Unlike his colleagues, however, Fuller focused in his considerations less on explicit linguistic models than on calculation and empiricism by means of mathematics and

---

41   Fuller 1975. See also Plato 2008, 53e, 54 a – c.

42   Snyder 1980, 115.

43   The inventor of the geodesic dome in lightweight construction is thought to be Walther Bauersfeld with his Zeiss Planetarium in Jena (1919 – 1926). Unlike Buckminster Fuller, however, Bauersfeld did not have his dome patented.

44   Krausse 2002, 42.

45   Rowe 1975.

46   Rorty 1967.

practical geometry. In his material-based transformation studies he discovered that the geometrical derivation of Platonic solids could be reduced, starting with a cuboctahedron via an icosahedron to an octahedron, and finally via the fold and inversion, to a triangle. This is the minimal, irreducible unit of every structurally relevant construction: "the triangle is a model in which each side stabilizes the opposite angle with a minimum of effort."[47] In space this corresponds to a tetrahedron. This principle thus replaces the traditional framework based on a square. As a so-called four-bar frame the latter was still used for example by the Stuttgart architect Frei Otto, who was likewise oriented to bionics. In 1972, as director of the Institut für freitragende Flächenbauwerke, Otto built the Olympic stadium in Munich with its spectacular roof.[48]

Buckminster Fuller thus categorically rejected the previous orientation to the square. With his triangles and by deploying projective geometry, he attempted to propose an alternative to the Cartesian coordinate system and to Euclidian space. With his focus on three-dimensional processes as opposed to planar geometry, Buckminster Fuller transferred the Greek idea of the close disposition of spatiotemporally bound forms – such as the *ichnographia* (plan), *orthographia* (elevation), and *scenographia* (perspective) – to a dynamically conceived system model. For this he used sketches in the form of vector plans, which did not, however, entail the disengagement from every kind of illusionism to the benefit of an idea of timelessness. Rather, Fuller drew on modern spatial concepts of physics, such as Erwin Schrödinger's space-time structure and Albert Einstein's search for a model for relativity theory.[49] For this, Buckminster Fuller coined the term synergetics, which he used to launch the idea of an energetic geometry.[50] From this he derived the idea of a structure model, with which as a side note he reformed crystallography.[51]

Buckminster Fuller did not, however, limit himself to the engagement with projective geometry; he also placed this in relation to a visual as well as tactile space of experience.[52] His embrace of the structural laws of nature in connection with an actively participating subject in dance marks him out as a realist. In other words: the spiritual participation of external realities shifts to the center of Fuller's aesthetic, which distinguishes him from

---

47   Snyder 1980, 117.
48   Nerdinger 2005.
49   Einstein 1923.
50   Snyder 1980, 135.
51   Fuller could demonstrate that there is a five-fold symmetry, which had previously been strictly rejected by representatives of crystallography. "It was Fuller's Expo-Dome, which Kroto and Smalley had seen in 1967 in Montreal, which posited them on the right track. Architecture as a catalyst of a scientific knowledge? Why not – especially when such a case of pattern recognition repeatedly occurred. The first catalysis of this type took place during 1959 – 62, as Donald Caspar and Aaron Klug [Nobel Prize in Chemistry, 1982] discussed with Fuller the structure of the viruscapsid." In: Krausse 2000, 211 (trans. by S.S.).
52   Buckminster Fuller called his series of experiments with Platonic solids, which he staged with one model, "five ways to dance the Jitterbug." Krausse 2002, 40 – 49, cf. also id. 1998, 293 – 294.

a nominalist, who is concerned with mere orders of nature. Accordingly, his classes at Black Mountain College in North Carolina did not foreground fixed typologies, but taught according to the principle of how nature itself builds. Unfortunately, the question of which nature is being referred to here is not fully answered. As is known, the concept of nature does not arise from a reception, but is the result of an active construction.[53] Buckminster Fuller's concern therefore was to convey a method of acquiring an empirically based structural knowledge. The logic of this was not exhausted in the mimetic imitation of natural structures; it also lay in the anticipation of the ideal forms of a practical artist-engineer.[54]

Renaissance of Pragmatism

If American functionalists in the first half of the 20th century focused on the idea of an evolutionary life process and an adaptation to the environment, which they tested out above all in experiments, then the protagonists of poststructuralism, Jacques Derrida and Gilles Deleuze among others, were concerned with a language-centered engagement with theory. This tendency, which arose out of linguistics, led to the establishment of the subjects of architectural history and theory at the ETH Zürich by Bernhard Hoesli in 1967 as well as at the Institute for Architecture and Urban Studies founded at the same time by Peter Eisenman in New York.[55] In this context, the launching of Eisenman's publications *Oppositions* and *Skyline* also made an essential contribution to an autonomization of architecture via the detour of language.[56] Subsequently, Eisenman developed the idea of architecture as an intertextual system, which in the 1980s he gradually expanded with concepts and methods derived from generative grammar after Chomsky, as well as from phenomenology and philosophy.[57] In this way, Eisenman can be seen as a direct successor to the English architectural theorist Colin Rowe, who in the immediate postwar period formulated an architecture as a self-referential system of signs characterized by trans-temporalities and structural thought, which permitted cross-references between Andrea Palladio and Le Corbusier. In this connection, Eisenman became among other things a model for the artist-architect Gordon Matta-Clark, whose projects tested new formats both on paper and in space.[58]

---

53 Haag 1973.
54 "I did not copy nature's structural patterns. I did not make arbitrary arrangements for superficial reasons. What really interests me therefore in all these recent geodesic tensegrity findings in nature is that they apparently confirm that I have found the coordinate mathematical system employed in nature's structuring. I began to explore structure and develop it in pure mathematical principle out of which the patterns emerged in pure principle and developed themselves in pure principle. I then realized those developed structural principles as physical forms, and in due course applied them to practical tasks." In: Fuller 1964, 59.
55 Frampton/Latour 1980, 5–41.
56 Eisenman 2004a, 227–231.
57 Eisenman 2005, 224.
58 Ursprung 2012, 38f.

Architecture of the Event

In the 1990s Eisenman turned his attention to an architecture of the philosophy of the event. This is characterized by a relation between inside and outside, as described by Deleuze in the figure of the rhizome.[59] Instead of fixed hierarchical attributions of positions in space and place, the rhizome takes up the tradition of Leibniz's theodicy.[60] In this context, with regard to the increase of complexities and manifolds as opposed to the essences in Plato, Joseph Vogl formulates the essential difference between Leibniz and Deleuze: "Rather than a question about the essence, we are presented with a polyphony of questions relating to the accident, to the manifold, and to the event."[61]

If Buckminster Fuller's tensegrity structures were still governed by the systematics of transformation on the basis of platonic solids, Eisenman follows in his transformations the logic of the philosophy of the event characterized by singularity. At the same time, with the shift from intertextuality to folding, one can discern a renaissance of pragmatism. However, unlike the pragmatists of the turn of the century who were concerned with biologistic models, Eisenman refers to a machine model, at whose basis, according to Horst Bredekamp, one can also discern the imitation of nature.[62]

It is not by chance therefore that Eisenman drew on mathematical models, especially those of René Thom, whose chaos theory deals with dynamic types. Here, Thom examined dynamic systems[63] with regard to their qualitative changes resulting from volatility.[64] As with all mathematical problems, Thom was concerned with the abstract modeling of the real world. More precisely, he was engaged with the description of the border zones of fractals in chaos theory. Here, mathematical catastrophe is described as an abrupt change of state leading to folds: "In the three-dimensional diagram of such a nonlinear, discontinuous system one recognizes in the critical threshold of a sudden change of state a folding of the diagram, which is also described as a catastrophe."[65]

In the mathematical model of chaos theory Thom distinguishes the forms cusp, swallowtail, butterfly, wigwam, elliptic umbilic, hyperbolic umbilic, parabolic umbilic, symbolic umbilic, and fold.[66]

---

59   Deleuze/Guattari 1987, 4–25.
60   Leibniz 2013.
61   Vogl 2003.
62   Bredekamp 1995.
63   Thom 1975, chapters 4, 5. Cf. also Aubin 2004, 95–130 and Aubin 2001, 255–279.
64   Zimmermann 1980, 385.
65   Werner 1996, 85.
66   Zimmermann 1980, 385–412.

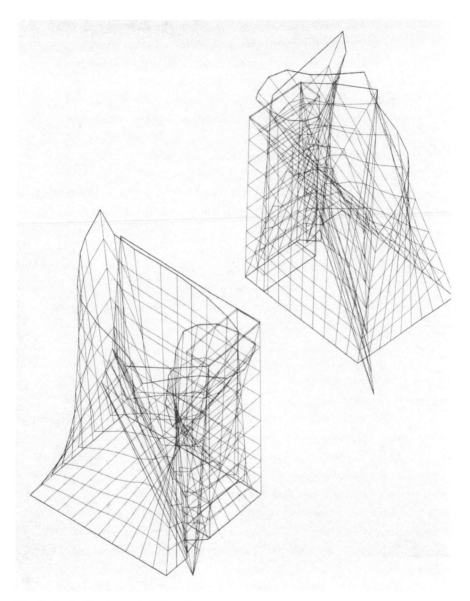

Fig. 4 Design for Haus Immendorff, Peter Eisenman.

As a source for his design for Haus Immendorff from 1993, Eisenman made use of Thom's model of the soliton wave, which represents an example of physical self-organization characterized by nonlinearity. The function of linearity is replaced here by curves and refractions. In this relation, Thom mentions both the Mandelbrot set and the Koch snowflake.

"The fold as a technique in architecture can accomplish opposite qualities: it can represent a sudden change of direction, assumption, or mood [...] Conversely it can resolve differences in a way which is distinct from the other architectural methods of dealing with pluralism, such as collage. This is by *enfolding,* by connecting that which is different in a smooth transition. Here suppleness and smoothness are important – the way, for instance, that two different liquids are enfolded into each other by stirring."[67]

In these formalizations folding is thus attributed the function of dynamic changes that are not only of use for purposes of communication, but also become transferrable entirely in the spirit of the modern sciences.[68] At this point it can also be surmised why Eisenman saw in chaos theory new possibilities for architecture.[69] With the use of mathematical models he promised himself a way out of the cul-de-sac of postmodernism, which while stimulating architectural discourse, tended formally to be geared to the combination and repetition of form. Against repetition, mathematical models of singular phenomena promised new approaches to architecture.

Forms of Representation

In modern mathematics once can distinguish among other things three different forms of representation, which Gert Höfner has characterized as follows:[70] analytic representation – that is, representation through formulae – enables the precise calculation of function values at the expense of the possibility of visualization; graphic representation – that is, representation in the Cartesian coordinate system – which has the advantage of the possibility of visualization, but the disadvantage of less exact values dependent on the increase of the area presented; finally, tabular representation or a table of values, which does not necessitate numerical calculation, but as a result need not display any values.[71] The missing links between mathematical exactness and visualization articulated in these models therefore no longer allows a transfer of the mathematical model between abstract formulae and real representation. This can be traced back to the founder of analytic geometry, René Descartes who provided the basis for the calculation of geometrical problems by forfeiting

---

67 Jencks 1995, 53f.

68 Eisenman 2007d.

69 Eisenman 2004b, 235f.

70 Höfner 2015.

71 Among other uses of models in modern mathematics, one should mention model theory, which was first developed in the 1930s, as well as physical models (made from card, string, plaster, and wood, etc.), which were abundant in Europe and in the USA, especially in the 19th century. Cf. Mehrtens 2004, 276–306.

the possibility of visualization.[72] Against this background, also the enforcement of the graphic representation of spatial structures can be designated as a model, which can also be described as a function diagram with three variables, and which function as orthonormal coordinate system. Complex dynamic systems are here designated as a folding of determining functions. Thus, differential geometry as a mathematical symbol system is able to replace the model of the Euclidean system. In this connection, the new spatial model of the rhizome also gains in importance, which takes the place of a centralized world model, and can be traced back to the beginnings of modern physics. At the beginning of the 20th century the mathematician Henri Poincaré explained the fundamental difference between what rests on measurement – that is, on the haptic – and projective Euclidean geometry.[73] This distinction brought about new approaches to geometry for use in architectural representation. In the 1980s the architectural theorist Robin Evans (1944–1993) referred to Poincaré, and saw the possibility of considering active as opposed to passive geometries as fields of force in the interstices of conventional gestalt theories. Evans remarked that geometrical composition loses its significance when it is not considered as a tool of projection, which he qualified as projection, quasi projection, and pseudo projection. According to Evans, the tools of these projection types provided new possibilities, as now the fourth, that is, temporal, dimension could be included in the architectural representation. Significant in this context are the so-called fields of force of drawing, which provide the key for the handling of geometries. Within this theory the fold assumes a paradigmatic meaning of a vector-derived geometry that cannot be sufficiently grasped using the figure–ground relation of gestalt theory.[74] In this logic, Eisenman demands for the representation of his folding architecture, beyond the typical architectural parallel projection of plan, elevation, and section, a topological model in which the index assumes the significance of the reference structure. Here, the index can indicate an unseen or virtual movement. Thus, Deleuze understands the index no longer as map with the reference to the grid, but as schema.

Perceptual Space

The aspect of reference at the basis of the schema reveals a further approach to the fold, which according to Deleuze can "read into [...] the soul."[75]

In turning away from the stylistic pluralism of postmodernism, Eisenman draws attention, besides the engagement with geometries of the fold, to an event, which in the Kantian sense lies "before the synthesis of perception."[76] In this way, perception becomes an elementary point

---

72   Descartes 1637.
73   Poincaré 1921.
74   Eisenman 2005, 226.
75   Deleuze 1993, 3.
76   Vogl 2003.

of reference in Eisenman's architecture: "perception alone dislocates the knowing subject."[77] Here, Eisenman understands space not as a precondition of perception in the Kantian sense, but as an atmosphere linked to time. Because in the neopragmatism of postwar architecture the end of theory was evoked and a greater autonomy and engagement with everyday culture was expected of architecture, one rediscovered theories of perception stretching back to the 18th century. Eisenman, alongside other theory-forming protagonists of architecture, can draw on the principles of gestalt theory and, with reference to an active geometry, demands its transgression in order to attain the category of a temporally determined atmospheric space:

> "[F]olded space articulates a new relationship between vertical and horizontal, figure and ground, inside and out – all structures articulated by traditional vision. Unlike the space of classical vision, the idea of folded space denies framing in favor of a temporal modulation. The fold no longer privileges planimetric projection; instead there is a variable curvature. [...] [F]olding [...], in terms of traditional vision, [...] contains a quality of the unseen. Folding changes the traditional space of vision. That is, it can be considered to be *effective*; it functions, it shelters, it is meaningful, it frames, it is aesthetic. Folding also constitutes a move from *effective* to *affective* space."[78]

Atmospheres

The inside–outside relation, which in Eisenman's architecture replaces the concept of the grid with that of the frame, becomes further charged in the context of theories of the event. With affective space, Eisenman wants to bring a modern consciousness to his formal justification.[79] Eisenman makes explicit reference to the architect Claude Perrault (1613–1688), who is considered Vitruvius's most important translator into French, and in the theoretical dispute with François Blondel demanded a rejection of an aesthetics of nature in favor of temporal concepts of beauty. Eisenman follows this tradition by extending the network of references in the design process beyond scientific standardization and timeless aesthetics to include the instabilities of chaos theory as well as references to fiction and psychoanalysis. In this way, temporal conditions come to the fore, which are not grasped with the concept of type, but rather with that of the topos.[80] With this approach Eisenman, drawing on phenomenology, also questions the duality of a subject–object relation as well as spatial preconditions in favor of experiences of the event. For Eisenman, this characterizes the modern consciousness.[81]

---

77 Eisenman 2007c, 41.
78 Eisenman 2005, 38.
79 Eisenman 2007a.
80 Eisenman 2004b, 236–237.
81 Schwarz 1995, 25.

Projective Space

Beyond this, the new consciousness is treated by Eisenman as a genuine architectural problem of representation. With a focus on a polymorphic perception, he projects the new consciousness onto a new space of representation. Drawing on Erwin Panofsky,[82] who in his theory treated perspective not as a mathematical but as a symbolic construction, Eisenman hoped from the fold for a new space of representation. Just as Evans questioned the cultural conventions of Durand's drawing systematics and situated drawing not only in relation to its episteme between drawing and calculating, but also claimed play as a quantity beyond calculability, Eisenman wants to explore the expressive possibilities of the polymorphic perception of form:

> "Whence the question: how can one return from this perception fash-
> ioned by culture to the 'brute' or 'wild' perception? What does the
> informing consist in? By what act does one undo it (return to the
> phenomenal, to the 'vertical' world, to lived experience)?"[83]

In this way, Eisenman acts in analogy to the painters of the time of Alberti, the so-called patron of projection – painters who missed in perspective the reflection of human perceptions and therefore categorically rejected the standards of drawing conventions, while in the architecture of the period the form of symmetry[84] and a correlation between architectural plan and built space prevailed.[85] Hierarchically centered symmetries following the logic of surveying in cartography have remained in use up to the present as a standard of architectural drawing. They refer back to Euclidean space, which is equated with the sense of touch. In the history of architectural theory, in contrast, the space of vision is associated with projective geometry. In the conflict between Euclidean and projective space Einstein defended Euclidean space by ascribing to it three subcategories: besides Euclidean, he also names Riemannian spherical and stereographic perspectival geometries. According to Evans, he thereby avoided the logic of duality by combining in his theory a field of anamorphic vision consisting of a balance of forces and a Euclidean metric. Within this theory he situated the sense of sight between geometry and visualization, between sensations, motor activities, and concepts.[86] In this way, perceptions and interpretations are raised to the highest principle, and the shaping of perception is opposed to a "petrified Platonism." Evans, who in his engagement with architectural drawing analyzes the advance of ideas about images, can therefore link geometry via

---

82 Panofsky 1991.
83 Maurice Merleau-Ponty quoted in Schwarz 1995, 11.
84 Philipp 2007, 147–157.
85 Eisenman 2007d.
86 Evans 1995, 352.

the representation of ideal forms in Greek geometry to the manipulation of ideal forms in the geometry of Descartes, and finally to modern models of knowledge.

Here projection becomes central for questions about constructions of the world, whose communication rests on the visualization of graphic representations of knowledge.

Thus, with modern physics conventional assumptions about representation are questioned.[87] Not representation but the affect placed before perception becomes the focus of the formalization of knowledge. If during the Renaissance a decisive role was played by relations of similarity in the framework of a creation theology, then Descartes translated these into that of a visualization strategy, although he was aware of the distinction between perception and information processing. Descartes differentiated between understanding and sensory perception.[88] Knowledge about the interactions between conceptual and perceptual space highlights the significance of the formalization of the design process for consciousness and vice versa: "Imagination is not held within the mind, but is potentially active in all the areas of transition from persons to objects or pictures. It operates, in other words, in the same zones as projection and its metaphors."[89]

Starting from a modern consciousness, which Eisenman refers to via the detour of philosophy, Euclidian space in the categories developed by Einstein seems only partly suitable for the construction of the fold. Irregular, singular displacements of the grid simultaneously lead to a topology, or put differently a correlation to the diagram, which through its interstices always also factors in the conceptual space of spatially ordered information. In the framework of semiotics, already Charles Sanders Peirce saw in the diagram the possibility of the visualization of thought in general guaranteed.[90]

The fold in Eisenman's architecture thus moves between Euclidean and topological diagrammatic geometry on the one hand, and affective space on the other. It arises as a consequence of the pluralism of postmodernism and is initially a development of neopragmatism, to finally fulfill, with the concept of affect, architecture's claim to autonomy and singularity. Hence, the fold is a form that on the basis of the grid arises through singular displacements, and is concerned not with combinations of forms as in postmodernism, but with experience-based multiperspectivity. In this way, the form of the fold brings scientific models to bear, without, however, relinquishing the space of singular perception that the event, which since Deleuze is unavoidably linked to the fold, first delivers.

---

87   Eisenman 1995, 582.
88   Descartes 1641, 21.
89   Evans 1995, 363.
90   Peirce 1983.

Bibliography

Adams, Robert M. (1998): *Leibniz: Determinist, Theist, Idealist*. Oxford: Oxford University Press.

Alberti, Leon Battista (1435): *Della Pittura. Über die Malkunst*. Ed. by Gianfreda, Sandra / Bätschmann, Oskar (Hg.). Darmstadt: Wissenschaftliche Buchgesellschaft 2002.

Alberti, Leon Battista (1454): *De re aedificatoria. Zehn Bücher über die Baukunst*. Ed. by Theuer, Max. Wien / Leipzig: Heller 1991.

Aubin, David (2001): *From Catastrophe to Chaos: The Modeling Practices of Applied Topologists*. In: Bottazzini, Umberto / Dahan-Dalmédico, Amy (eds.): Changing Images in Mathematics: From the French Revolution to the New Millennium. London: Routledge, pp. 255–279.

Aubin, David (2004): *Forms of Explanation in the Catastrophe Theory of Rene Thom: Topology, Morphogenesis, and Structuralism*. In: Wise, M. Norton (ed.): Growing Explanations: Historical Perspective on the Sciences of Complexity. Durham: Duke University Press, pp. 95–130.

Böhme, Gernot (2013): *Atmosphäre: Essays zur neuen Ästhetik*. Berlin: Suhrkamp Verlag.

Bredekamp, Horst (1995): *The Lure of Antiquity and the Cult of the Machine: The Kunstkammer and the Evolution of Nature, Art and Technology*. Princeton: M. Wiener Publishers.

Bredekamp, Horst (2004): *Die Fenster der Monade: Gottfried Wilhelm Leibniz' Theater der Natur und Kunst*. Berlin: Akademie Verlag.

Breidbach, Olaf (1998): *Ernst Haeckel: Kunstformen der Natur*. Munich: Prestel.

Cassirer, Ernst (1956): *Die Begriffsform im mythischen Denken*. In: id.: Wesen und Wirkung des Symbolbegriffs. Darmstadt: Wissenschaftliche Buchgesellschaft, pp. 1–70.

Deleuze, Gilles (1993): *The Fold: Leibniz and the Baroque*. Trans. by Conley, Tom. London: The Athlone Press.

Deleuze, Gilles / Guattari, Félix (1987): *Introduction: Rhizome*. In: id.: A thousand Plateaus: Capitalism and Schizophrenia. Trans. by Massumi, Brian. Minneapolis / London: University of Minnesota Press, pp. 1–25.

Derrida, Jacques (1987): *The Truth in Painting*. Trans. by Bennington, Geoffrey / McLeod, Ian. Chicago: The University of Chicago Press.

Descartes, René (1637): *La Géométrie*. Dover: Dover Publications 2012.

Descartes, René (1641): *Meditations on First Philosophy with Selections from the Objections and Replies*. Ed. by: Cottingham, John. Cambridge et al.: Cambridge University Press 2015.

Durand, Jean Nicolas Louis (2000): *Précis of the Lectures on Architecture with Graphic Portion of the Lectures on Architecture*. Los Angeles, CA: Getty Research Institute.

Dürer, Albrecht (1538): *Underweysung der Messung, mit dem Zirckel un richtscheyt, in Linien, Ebnen und gantzen Corporen*. Nuremberg: Hieronymus formschneyder.

Einstein, Albert (1923): *The Meaning of Relativity: Four Lectures Delivered at Princeton University, May, 1921*. Princeton: Princeton University Press.

Eisenman, Peter (1995): *Aura und Exzeß: Zur Überwindung der Metaphysik der Architektur*. Ed. by Schwarz, Ullrich. Vienna: Passagen Verlag.

Eisenman, Peter (2004a): *Architecture as a Second Language: The Texts of Between*. In: id.: Inside Out: Selected Writings, 1963–1998. New Haven / London: Yale University Press, pp. 226–233.

Eisenman, Peter (2004b): *Blue Line Text*. In: id.: Inside Out: Selected Writings, 1963–1998. New Haven / London: Yale University Press, pp. 234–237.

Eisenman, Peter (2005): *Ins Leere geschrieben: Schriften & Interviews 2*. Ed. by Engelmann, Peter. Vienna: Passagen Verlag.

Eisenman, Peter (2007a): *The Affects of Singularity*. In: id.: Written into the Void. New Haven / London: Yale University Press, pp. 19–24.

Eisenman, Peter (2007b): *Presentness and the Being-Only-Once of Architecture*. In: id.: Written into the Void. New Haven / London: Yale University Press, pp. 42 – 49.

Eisenman, Peter (2007c): *Visions Unfolding: Architecture in the Age of Electronic Media*. In: id.: Written into the Void. New Haven / London: Yale University Press, pp. 34 – 41.

Eisenman, Peter (2007d): *Processes of the Interstitial: Notes on Zaera-Polo's Idea of the Machinic*. In: Written into the Void. New Haven / London: Yale University Press, pp. 50 – 71.

Evans, Robin (1995): *The Projective Cast: Architecture and its Three Geometries*. Cambridge, MA / London: The MIT Press.

Fludd, Robert (1619): *Microscosmi historia. Tomus secundus de supernaturali, naturali, praeternaturali et contranaturali microcosmi historia intractatus tres distributa*. Oppenheim: Johann Theodor de Bry.

Frampton, Kenneth / Latour, Alessandra (1980): *Notes on American Architectural Education*. In: Lotus International, vol. 27, no. 2, pp. 5 – 39.

Fuller, Richard Buckminster (1964): *World Design Initiative*. In: id.: World Design Science Decade, phase 1, document 2, The design initiative. Ed. by: McHale, John. Carbondale, Illinois: World Resources Inventory, Southern Illinois Univ., pp. 1 – 104.

Fuller, Richard Buckminster (1982): *Synergetics: Explorations in the Geometry of Thinking*. New York: Macmillan.

Galilei, Galileo (1610): *Die Vermessung der Hölle*. In: id.: Sidereus Nuncius, Nachricht von neuen Sternen. Ed. by Blumenberg, Hans. Frankfurt a. M.: Suhrkamp Verlag 1980, pp. 231 – 250.

Goldman, Eric F. (1997): *Rendezvous with Destiny: A History of Modern American Reform*. New York: Vintage Books.

Grelon, André (1994): *Von den Ingenieuren des Königs zu den Technologien des 21. Jahrhunderts: Die Ausbildung der Ingenieure in Frankreich*. In: id. (ed.): Ingenieure in Frankreich, 1747 – 1990. Frankfurt a. M.: Campus, pp. 15 – 57.

Haag, Karlheinz (1973): *Der Fortschritt in der Philosophie*. Frankfurt a. M.: Suhrkamp Verlag.

Haeckel, Ernst (1913): *Die Natur als Künstlerin*. Ed. by Goerke, Franz. Berlin: Vita Deutsches Verlagshaus.

Hoffmann, Volker (1992): *Filippo Brunelleschi: Kuppelbau und Perspektive*. In: Bozzoni, Corrado / Carbonara, Giovanni / Villetti, Gabriella (eds.): Saggi in onore di Renato Bonelli: Quaderni dell'istituto di storia dell'architettura. Rome: Multigrafica Editrice, pp. 317 – 327.

Höfner, Gert (2015): *Mathematik im Millimeterpapier*. Letter from July 20.

Hutterer, Robert (1998): *Das Paradigma der Humanistischen Psychologie: Entwicklung, Ideengeschichte und Produktivität*. Vienna / New York: Springer Verlag.

Jencks, Charles (1995): *The Architecture of the Jumping Universe*. New York: Academy Editions.

Kant, Immanuel (1786): *Metaphysische Anfangsgründe der Naturwissenschaft*. Riga: Hartknoch.

Koyré, Alexandre (1990): *From the Closed World to the Infinite Universe*. Baltimore / London: The Johns Hopkins Press.

Krausse, Joachim (1998): *Buckminster Fullers Vorschule*. In: Fuller, Richard Buckminster: Bedienungsanleitung für das Raumschiff Erde und andere Schriften. Ed. by Krausse, Joachim. Amsterdam / Dresden: Verlag der Kunst, pp. 213 – 306.

Krausse, Joachim (2000): *Das Zwinkern der Winkel*. In: Fecht, Tom / Kamper, Dietmar (eds.): Umzug ins Offene: Vier Versuche über den Raum. Vienna / New York: Springer, pp. 187 – 214.

Krausse, Joachim (2002): *Buckminster Fullers Modellierung der Natur*. In: Arch+, vol. 159 – 160, pp. 40 – 49.

Leibniz, Gottfried Wilhelm (2013): *Philosophische Schriften: Die Theodizee von der Güte Gottes, der Freiheit des Menschen und dem Ursprung des Übels*, vol. 2. Ed. by Herring, Herbert. Darmstadt: WBG.

Leibniz, Gottfried Wilhelm (1978): *Die philosophischen Schriften*. Ed. by Gerhardt, Carl Immanuel, vols. I – VII. Berlin: Olms.

Maak, Niklas (2010): *Architekt Claude Parent: Auf der schiefen Bahn*. In: Frankfurter Allgemeine Sonntagszeitung, no. 30, 1 August, p. 20.

Mehrtens, Herbert (2004): *Mathematical Models*. In: de Chadarevian, Soraya / Hopwood, Nick (eds.): Models: The Third Dimension of Science. Stanford, CA: Stanford University Press, pp. 276–306.

Nerdinger, Winfried (ed.) (2005): *Frei Otto: Complete Works: Lightweight Construction, Natural Design*. Basel: Birkhäuser Verlag.

Panofsky, Erwin (1991): *Perspective as Symbolic Form*. Trans. by Wood, Christopher S. New York: Zone Books.

Peirce, Charles S. (1983): *Phänomen und Logik der Zeichen*. Ed. and trans. by Pape, Helmut. Frankfurt a. M.: Suhrkamp.

Philipp, Klaus Jan (2007): *Die Imagination des Realen: Eine kurze Geschichte der Architekturzeichnung*. In: Gleiter, Jörg / Korrek, Norbert / Zimmermann, Gerd (eds.): Die Realität des Imaginären: Architektur und das digitale Bild (10. Internationales Bauhaus-Kolloquium Weimar). Weimar: Schriften der Bauhaus-Universität Weimar, pp. 147–157.

Plato (2008): *Timaeus*. In: id.: Timaeus and Critias. Trans. by Waterfield, Robin. Oxford: Oxford University Press, pp. 1–100.

Poincaré, Henri (1921): *Des fondements de la géométrie*. Paris: Etienne Chiron.

Rajchman, John (1998): *Folding*. In: id.: Constructions. Cambridge, MA / London: The MIT Press, pp. 11–36.

Riedler, Alois (1896): *Das Maschinen-Zeichnen: Begründung und Veranschaulichung der sachlich notwendigen zeichnerischen Darstellungen und ihres Zusammenhangs mit der praktischen Ausführung*. Berlin: Springer.

Rorty, Richard M. (1967): *The Linguistic Turn: Essays in Philosophical Method*. Chicago: The University of Chicago Press.

Rowe, Colin (1975): *Introduction*. In: Eisenman, Peter (ed.): Five Architects: Eisenman, Graves, Gwathmey, Hejduk, Meier. New York: Oxford University Press, pp. 3–8.

Schwarz, Ullrich (1995): *Another look – anOther gaze: Zur Architekturtheorie Peter Eisenmans*. In: Eisenman, Peter: Aura und Exzeß: Zur Überwindung der Metaphysik der Architektur. Ed. by Schwarz, Ullrich. Vienna: Passagen Verlag, pp. 11–34.

Sennett, Richard (1994): *Flesh and Stone: The Body and the City in Western Civilization*. London: Faber and Faber.

Snelson, Kenneth: *What is Tensegrity?* Online: *www. kennethsnelson.net / faqs / faq.htm* (last access: 17 December, 2015).

Snyder, Robert (1980): *Buckminster Fuller: An Autobiographical Monologue/Scenario*. New York: St. Martin's.

Spencer, John R. (ed.) (1965): *Filarete's Treatise on Architecture. Volume 1: The Text*. New Haven / Connecticut / London: Yale University Press.

Strickland, Lloyd (ed.) (2014): *Leibniz's Monadology: A New Translation and Guide*. Edinburgh: Edinburgh University Press.

Thom, René (1975): *Structural Stability and Morphogenesis: An Outline of a General Theory of Models*. Trans. by Fowler, David H.. Reading et al.: W. A. Benjamin.

Thompson, D'Arcy Wentworth (1917): *On Growth and Form*. Cambridge: Cambridge University Press.

Ursprung, Philip (2012): *Grenzen der Architektur*. In: Von Amelunxen, Huberts / Lammert, Angela / Ursprung, Philip (eds.): Gordon Matta-Clark: Moment to Moment: Space. Berlin: Verlag für Moderne Kunst, pp. 29–47.

Virilio, Paul (1997): *The Oblique Function*. In: Virilio, Paul / Parent, Claude: Architecture Principe: 1966 and 1996. Trans. by Collins, George. Besançon: Les Éditions de l'Imprimeur, pp. iii–v.

Vogl, Joseph (2003): *Was ist ein Ereignis?* Lecture given at the Center for Art and Media (ZKM). Karlsruhe, 26 October. Online: *http://zkm.de/media/ audio/joseph-voglwas-ist-ein-ereignis* (last access: 25 January 2016).

Werner, Hilmar (1996): *Falten im Leichtbau*. In: Arch+, vol. 131, pp. 82–85.

Zimmermann, Rainer (1980): **Katastrophentheorie
und die Geometrie der Entscheidungen**. In: Blätter
der DGVFM, vol. 14, no. 3, pp. 385–412.

Zur Lippe, Rudolf (1974): **Naturbeherrschung am
Menschen**, 2 vols. Frankfurt a. M.: Suhrkamp Verlag.

Sandra Schramke
Email: *sandra.schramke@hu-berlin.de*

Image Knowledge Gestaltung. An Interdisciplinary Laboratory.
Cluster of Excellence Humboldt-Universität zu Berlin.
Sophienstrasse 22a, 10178 Berlin, Germany.

Michael Friedman, Joachim Krausse

Folding and Geometry: Buckminster Fuller's Provocation of Thinking

1. Introduction

From the 19th century to the early 20th century, geometry had changed its character considerably. Discoveries, such as non-Euclidean geometry, alongside the development of differential geometry with its definition of the manifold, instigated a plurality of geometries. Encompassing this plurality was a trend of thought that called for situating geometry on stable foundations, one might say even static pre-determined ones. Against this backdrop of a growing move towards axiomatization, Richard Buckminster "Bucky" Fuller (1895–1983), an American architect, designer and inventor, offered a critic of Euclidean and Cartesian geometry from a novel reconsideration of practical actions and operations like folding – folding, as a form of thinking on and through movement, enabling a different conception of geometry. This paper aims to show that beyond an axiomatized motionless geometry, on the one hand, and the various forgotten mathematizations of the fold, on the other, Fuller suggests to think of movement from a different perspective: movement as the provocation of thinking. It is what provokes and initiates thinking itself. Starting with Fuller's critique of geometry and concluding with his conception of mobility, we examine notions of movement present in Fuller's thought. Indeed, folds and folding lie at the core of Fuller's work as an example of mastering movement.

2. Fuller and Geometry: Fuller's critique and the conception of geometry and folding at the beginning of the 20th century

Before turning to Fuller's conceptions of the fold and mobility, as what provokes stable structures and buildings, we examine Fuller's critique of the axiomatic conception of geometry, as exemplified in Euclidean axiomatics. We then review the manner in which geometry in general and folding in particular were perceived within mathematics, from the beginning of the 19th century till the middle of the 20th century, in order to assess correctly Fuller's critique of the problematic relation between movement and geometry and his conception of folding.

2.1 Fuller and the Euclidean geometry

Needles to say, Euclid's geometry – as presented in his book *Elements* – is one of the most influential theories of western civilization. However, little is known about the author, beyond the fact that he lived in Alexandria around 300 BCE. Most of the theorems appearing in the *Elements* were not discovered by Euclid himself, but were the work of earlier Greek mathematicians such as mathematicians of the Pythagorean School, Hippocrates of Chios, Theaetetus of Athens and Eudoxus of Cnidos. Credited to Euclid is the arrangement of these theorems in a logical manner, in order to show that they necessarily follow from basic definitions, postulates and axioms.[1] The geometrical constructions employed in the *Elements* are restricted to those achieved by using a straightedge and a compass. Empirical proofs using measurement were not allowed: i.e., the only statements that were allowed were these in form of declaring that magnitudes are either equal, or that one is greater than the other.

Euclid's rigor and organization was admired throughout the ages and considered as one of the main methods of proper mathematical investigation. What constitutes rigor has changed over the years: modern mathematics returned to Euclidean geometry, revealing missing axioms and finding gaps in proofs, while trying at the same time to reaffirm its consistency together with the consistency of the 19th century analytic geometry. Nevertheless, the basic tools and methods of Euclidean geometry persisted throughout the centuries: an infinite line, a circle and a scribe – a system of basic signs and propositions – from which every other true proposition can be derived.

It is at this point that Fuller attacks Euclid's geometry, by criticizing its tools: "Euclid limited himself in his theorems to construction and proof by the use of three tools –

---

1   Cf. Heath 1921, 319; Proclus 1992, 53.

straightedge, dividers, and scribe. He, however, employed a fourth tool without accrediting it – this was the surface upon which he inscribed his diagrammatic constructions."[2]

In his paper from January 1944, Fuller presents his position, in which he sets the groundwork for his "energetic geometry" that would later become "synergetics." Fuller follows up his critique, of the lack of use of a tool that was completely forgotten, by giving a historical explanation:

> "It must be remembered that Euclid argued his geometric cases at a time in history when the spherical concept of the universe, which some assert was known to ancient Greek philosophers, had if so, been lost again. At that time, the savants were subscribing to a flat or planar earth concept. Therefore, it is not surprising that his use of that flat plane as a surface upon which to work went as axiomatic. Logical to the misconception was the beginning of his proofs in the special abstract realm of an imaginary plane geometry."

Fuller's critique is in effect a contrarian stance against Euclid, whom he accuses of being the one who "had come in by the wrong entrance" and hence had insufficiently reflected upon his own tools. This has led, according to Fuller, to an illusory elementarism in the sciences: it not only reduced geometry into a sequence of logical steps, from which one could eventually draw a conclusion, but at the same time expelled from geometry the pivotal concept of movement,[3] at best reducing it to a secondary concept derived from more fundamental objects, which could be removed at any point from the geometrical structure of which it stems. Euclidean geometry, according to Fuller's conception, is static; the concept of movement is invoked through axioms, a step that can be avoided and is in fact redundant. Fuller says so explicitly, when he remarks:

> "We find experimentally that two lines cannot go through the same point at the same time. One can cross over or be superimposed upon

---

2   This citatation and the following two are taken from a 1944 paper by Fuller: *Dymaxion comprehensive system, introducing energetic geometry*. In: Krausse/Lichtenstein 2001, 160–168, here: 164.

3   The history of the mathematical geometrical use of the notions of motion and movement (for example, whether they should be used as tools in mathematical proofs, how they should be conceptualized, what kind of entities – curves, surfaces – do moving objects create) starts already in antiquity; it is intricate and subtle. Aristotle condemned the use of motion in Geometry, stating "[t]he objects of mathematics are without motion" (Aristotle 1928–1952, vol. 8, 989b), whereas Euclid does use the concept of motion in some of his definitions (Book XI of Euclid's *Elements*, definitions 14,18 and 21. See Heath 1908b, 261–262). For overviews concerning motion, space and geometry, see e.g. Rosenfeld 1988, esp. chapter 3 and De Risi 2015. As we merely aim to point at the mathematical background against which Fuller developed his own thought, we by no means attempt to give even a partial account of it, as it would take us outside the scope of this paper.

another. Both Euclidian and non-Euclidian geometries misassume that a plurality of lines can go through the same point at the same time. But we find experimentally that two or more lines cannot physically go through the same point at the same time."[4]

All known geometries presuppose the totality of all lines already exists, since only then can two lines pass through a single point *at the same time*. Fuller makes the claim geometry does not take into account the dimension of time, and therefore may also not take into account time-consuming movement required in order to draw a line.[5] The movement, which acts as the dynamic aspect of the structure, is in effect what keeps a built structure stable, as we will see in Section 3.1 in connection with Semper. According to Fuller, this is not apparent as long as one restricts oneself to plane geometry:

> "[...] the Greek geometers were first preoccupied with only plane geom-
> etry. They were also either ignorant of – or deliberately overlooked – the
> systematically associative minimal complex of *inter-self-stabilizing forces*
> (vectors) operative in structuring any system (let alone our planet)
> and of the corresponding cosmic forces (vectors) acting locally upon
> a structural system. These forces must be locally coped with to insure
> the local system's structural integrity [...]"[6]

It is clear Fuller's critique did not merely target Euclidean geometry as embedded in its context of origin. It was rather aimed at its revival during the late 19th century. It is here that we should take a step back in order to understand the mathematical landscape that served background to his critique. What was the conception of geometry during the end of the 19th century to the beginning of the 20th century? How were the concepts of motion and movement reshaped?

2.2 The structural understanding of geometry at the beginning of the 20th century

In this section we will briefly review the conception of geometry from the end of the 19th century until the middle of the 20th century, focusing on Felix Klein's *Erlangen program* and David Hilbert's *Grundlagen der Geometrie*, and finishing with Alfred Tarski's axiomatization of geometry. We wish to highlight that Fuller's critique did not solely take aim at Euclid's *Elements*; it was particularly interested in the revival of interest in axiomatic methods. At the end of the 19th century the interest in the foundations of geometry was growing

---

4   Fuller 1975a, section 517.03.
5   Hence, there is only a partial overlapping of events. See Section 3.2.
6   Ibid, section 986.042.

both from a group-theoretic viewpoint and an axiomatic viewpoint. The emergence of non-Euclidean geometry at the beginning of the 19th century (Bolyai's and Lobachevsky's treatises), the mathematization of space via Riemannian manifolds and the mathematical definition of curvature prompted major philosophical questions regarding the nature of space and its epistemology.[7] The emergence of non-intuitive geometries gave rise to a need to discover the relations between the axioms of geometry and experience. In order to give a proper albeit incomplete historical account of this period, we begin with a recourse to *group theory*, which was one of the main topics of mathematical investigation during the 19th century, and which served as one important source for the development of a conception of geometry of that time.

A *group*, denoted by the letter '*G*', is set of elements equipped with a binary action, denoted by '*', which fulfills certain requirements. An obvious example for a group is the set of whole numbers together with addition as its binary action. The requirements the action should fulfill are considered to be the most elementary, when we think about actions such as addition or multiplication. To be more specific, there are four requirements: *closure* (if the elements a,b belong to G, denoted as $a,b \in G$, then $a*b$ belongs to G, denoted as $a*b \in G$), associativity (if $a,b,c \in G$, then $a*(b*c) = (a*b)*c$), *unit element* (there exists an element $e \in G$ s.t. $e*g = g*e = g$ for every element $g \in G$) and *inverse element* (for every $g \in G$ there exists an $h \in G$ such that $g*h = h*g = e$).[8]

The study of group theory and its applications is usually considered to originate from the work of Évariste Galois (1811–1832), who was working on the necessary conditions for solving an algebraic equation using the four known arithmetical operations (addition, subtraction, multiplication and division) together with roots of any order. What interested Galois around 1830 was not the equations themselves or their solutions, nor was he interested in the type of algebraic relations the roots hold among themselves. He was interested instead in the set of permutations of the roots themselves that preserve their algebraic relations.[9] In other words, Galois's discoveries prompted a process by which numbers were no longer considered fundamental to algebra. More crucial was a grasp of the algebraic-structural setting for which numbers assembled into various sets serve only an example and considered as a derivative of this structure.

---

7   For an extensive survey on the changing face of geometry during the 19th century see Gray 2006.

8   This definition can be found in all textbooks on group theory. See e.g. Rotman 1999, 12.

9   For example, for the equation $x^4 - 5x^2 + 6 = 0$, the solutions are A = $\sqrt{2}$, B = $-\sqrt{2}$, C = $\sqrt{3}$, D = $-\sqrt{3}$ and one of their mutual relations is: AB+CD = -5. Not every permutation of the roots A, B, C and D will preserve this relation. For example, if the permutation, denoted by f, is A→B, B→C, C→D, D→A then f(A)f(B)+f(C)f(D) = BC+DA = $-2\sqrt{6} \neq -5$. More surprisingly, out of the set of 24 possible permutations of 4 elements, only 4 permutations preserve the above relation.

The concept of a *permutation group* was derived from developments in the theory of algebraic equations and from what became known as Galois theory. This historical strand is just one of the roots of group theory. Indeed, Klein's Erlangen program makes it clear the development of the concept of the abstract group had another historical root, namely, geometry. Felix Klein (1849–1925) was a German mathematician and mathematics educator, known for his work in group theory and non-Euclidean geometry, and for his work on the connections between geometry and group theory that spurred his Erlangen program.[10]

Klein's program incorporated the idea that to every geometrical entity one can associate an underlying group of symmetries. By *symmetry* we mean a one-to-one transformation of the space onto itself that preserves certain properties of the space in question. The notion of a group is essential here: its set of elements was the set of symmetries, and the binary action was composition, as in the composition of functions. If S is our space (e.g. S is the three-dimensional Euclidean space), and f is a symmetry transformation of S (e.g. f acts by rotation with respect to an axis) then there are distinct subsets of S, which are not transformed by f (e.g. the axis of rotation). From this standpoint, Klein stated the task of geometry as follows:

> "Given a manifold and a group of transformations of the manifold,
> to study the manifold configurations with respect to those features,
> which are not altered by the transformations of the group."[11]

The mathematical hierarchy of geometries is thus represented as a hierarchy of these groups, and the hierarchy of their invariants. For example, lengths, angles and areas are preserved with respect to the Euclidean group of two-dimensional symmetries, while only incidence and cross-ratio are preserved under the more general group of two-dimensional *projective transformations*. One might be under the impression that, in opposition to Fuller's conception of the Euclidean *Elements*, Klein's Erlangen program does indeed deal with movements and transformations (such as rotation, translation and reflection). However, let us consider the following citation from Klein's: "*We peel off the mathematically inessential physical image* and see in space only an extended manifold; [...] transformations of manifold [...] also form groups".[12] Together with peeling off the "inessential physical image", one obtains a removal of any physical movement at the foundation of geometry. In this respect, Fuller might have regarded Klein's program as a

---

10   Klein 1872.
11   Klein 1893, 67.
12   Ibid.

descendant of the axiomatic method: group theory deals with movement as an abstract movement that can and should be formalized and axiomatized; a static structure, that is.[13]

One consequence of Klein's program was that it enabled the acceptance of Hilbert's axiomatic-structural approach to geometry. Indeed, as Wussing states "the transition to the notion of an abstract group was a partial cause, as well as a partial effect, for the growing acceptance of the 'axiomatic method' in Hilbert's sense of the term."[14]

Recognized as one of the most influential mathematicians of the late 19th and early 20th centuries, David Hilbert (1863–1943) was a German mathematician, who advanced research on the axiomatization of geometry, culminating in one of his most influential works: *Grundlagen der Geometrie*. It should be noted that Hilbert was not the first to suggest geometry should return to its axiomatic origins. Moritz Pasch, Mario Pieri and Hermann Wiener,[15] among others, also dealt with the subject at that time. However, Hilbert's approach was decisive for the way geometry was conceived in the early 20th century. Hilbert conceived of geometry as a natural science, one in which intuition plays a crucial role, though its experimental foundations may be regarded somewhat retroactively.[16] Hilbert states in his lectures on mechanics:

> "Geometry is an experimental science [...]. But its experimental founda-
> tions are so irrefutably and so generally acknowledged, they have been
> confirmed to such a degree, that no further proof of them is deemed
> necessary. Moreover, all that is needed *is to derive these foundations
> from a minimal set of independent axioms* and thus to construct the
> whole edifice of geometry by purely logical means."[17]

Once a minimal set of independent axioms is put together, geometry is studied through logical means:

> "Geometry [...] requires for its logical development only a small number
> of simple, fundamental principles. [...] [T]he choice of the axioms and
> the investigation of their relations to one another is [...] tantamount to
> the logical analysis of our intuition of space. The following investigation

---

13   See Wussing 1984, Part III.2 for an extensive analysis of Klein's program, and 194–196 for a description
     of mechanical movements in terms of group-theoretic concepts. It should be noted that Klein was also
     an ardent supporter of the use of models in mathematical teaching and research especially in the field of
     geometry. See for example: Mehrtens 2004; Sattelmacher 2013; Rowe 2013.
14   Wussing 1984, 251.
15   Pasch 1882; Wiener 1892; Pieri 1898.
16   See Corry 2004, chapter 3.
17   Ibid, 162. See also Corry 1997.

is a new attempt to choose for geometry a simple and complete set
of independent axioms [...]"[18]

Hilbert's views on geometry in particular and mathematics in general therefore did not regard mathematics as an empty formal game;[19] they rather emphasized independence and consistency of an axiomatic system derived from intuition and experience. That view was promoted in *Grundlagen der Geometrie*, where Hilbert's objective was to identify and fill 'gaps' or remove 'extraneous hypotheses' in Euclid's reasoning. The manuscript laid out a clear and precise set of axioms for Euclidean geometry, and demonstrated in detail the relations of those axioms to one another and to some of the fundamental theorems of geometry.

In *Grundlagen der Geometrie* Hilbert considers three collections of basic objects, which he calls 'points', 'straight lines' and 'planes', and five relations between them. The conditions prescribed in Hilbert's system of axioms are sufficient to characterize the basic objects and their relation to each other. In order to prove axiomatic independence, Hilbert builds several different geometries by negating some axioms while keeping others intact. Albeit possibly counter-intuitive, the resulting geometries are consistent. Geometry's innate structure is maintained as a consistent one, unrelated to physical reality, to which it does not correspond. This can be seen in Hilbert's words:

> "We think of these points, straight lines, and planes as having certain
> mutual relations, which we indicate by means of such words as 'are
> situated,' 'between,' 'parallel,' 'congruent,' 'continuous,' etc."[20]

What points, lines and planes have are their relations to each other. An object 'point' does not necessarily refer to a point in the physical sense: the only necessary and sufficient condition for it to be such is that it satisfies the relations between what is called 'point', 'line' and 'plane'. It divorces geometry from any recourse to a specific instinctive meaning (or notions such as movement or motion). This was apparent already in 1893, when Hilbert, upon his return from Halle after hearing Wiener's lecture, famously said: "One should always be able to say, instead of 'points, lines, and planes', 'tables, chairs, and beer mugs'".[21]

The understanding that geometry is not about describing a space, but rather about conceiving it as what is grounded in a system of axioms, gave rise to a plurality of different

---

18   Hilbert 1899, 1.
19   See Corry 2004, 161.
20   Ibid, 2.
21   Blumenthal 1935, 402–3.

geometries. It opened the way to view geometry (and algebra) first and foremost as an internal structure, one that is not based on movement, measuring or counting. This is manifest in the shift from Hilbert to Tarski. Hilbert, having advanced mathematical formalism considerably, still regarded geometry as fundamentally empirical, though experimentation in itself need not be performed.[22] Tarski, on the other hand, considered geometry wholly in its structural interiority. Alfred Tarski (1901–1983) was a Polish logician, mathematician and philosopher considered as one of the greatest logicians of the 20th century. He proved in 1930 that geometry, once formulated according to a specific choice of notations and axioms, admits an elimination of quantifiers: every formula is equivalent to a Boolean combination of basic formulae, that is, geometrical propositions can be written using first order logic alone. Once setting up the basic objects, relations and axioms, every claim of Euclidean geometry can be formulated using the quantifiers ∃ ('there is') and ∀ ('for every') together with its basic objects serving as variables.[23]

While Hilbert is considered one of the influencing mathematicians to reformulate to Euclid's axiomatic geometry, it is Tarski who found a more economic and efficient axiomatization for it.[24] Tarski's system of axioms for Euclidean geometry was based on a single primitive element – 'point' – and two undefined relations among those elements – *betweenness* and *equidistance* (or congruence). For every three points a, b and c, the relation 'betweenness' takes the value 'true' if the point b lies on the line segment with ends a and c. For two pairs of points – thinking of each pair as the endpoints of a line segment – the relation 'equidistance' holds if the two segments are of equal length. All other relations are consequently derived; for example, the collinearity of three points is defined in terms of betweenness (a, b and c are collinear if and only if one of them is between the other two). Tarski did not take 'line' or 'incidence' to be primitive notions; indeed, the only primitive notion is the point.

The primary significance of Tarski's elementary geometry lies in its satisfying three essential meta-mathematical properties: it is *deductively complete* (every assertion is either provable or refutable), *decidable* (there is a procedure for determining whether or not any given assertion is provable), and it is *consistent* (and this is why it is a correct axiomatization).[25] In order to prove these three, Tarski, in a move similar to Hilbert's, based geometry on the real numbers. To prove the completeness of the systems of complex algebra and Euclidean geometry, Tarski proved the completeness of the system

---

22 Concerning Hilbert's contribution to the rise of modern algebra and modern geometry, see for example: Corry 2004, chapter 3; Mancosu 1998, Part III; Hasse 1932.

23 Here is an example of one claim of Euclidean geometry: for any triangle, the sum of the lengths of any two sides must be greater than or equal to the length of the remaining side.

24 By "more efficient" we mean that Tarski proved with this axiomatization that the euclidean geometry is consistent. For an extensive survey on Tarski's life and work, see Feferman/Feferman Burdman 2004.

25 See Tarski 1967.

of algebra based on real numbers – one that Hilbert assumed as evident and therefore did not bother to prove.[26] Not only that, Tarski noted: "it is possible to construct a machine which would provide the solution of every problem in elementary algebra and geometry".[27]

This mechanical description of geometry is expressed in Tarski's formulation: all axioms and propositions are expressed in terms of first order logic. For example, the famous parallel axiom can be expressed as follows:[28]

$$[B(abf) \wedge ab \equiv bf \wedge B(ade) \wedge ad \equiv de \wedge B(bdc) \wedge bd \equiv dc] \rightarrow bc \equiv fe$$

where the variables are points and $B(-,-,-)$ designates betweenness. This is a description that does not resemble Euclid's in any form: "If a line segment intersects two straight lines forming two interior angles on the same side that sum to less than two right angles, then the two lines, if extended indefinitely, meet on that side on which the angles sum to less than two right angles."[29]

In Tarski's framework one does not need *several* basic objects. Such plurality might induce problematic relations between these objects, or a tacit form of abuse of notation might take place, as seen in Hilbert's *Grundlagen der Geometrie*.[30] A single object is all that is called for – an abstract object without presupposed properties, bearing no particular relation to empirical reality or intuition.[31] Its properties are exclusively derived from a system of axioms: the point in Tarski's work is an object defined according to what satisfies the axioms.

What is then the essence of geometry in its various faces from Klein to Tarski? It is clear Fuller's critique bears merit, though not entirely well grounded from a historical stand-point. From Fuller's perspective, motion and movement were formulized so that they became pure mathematical objects, a maneuver that leads to a reduction of dynamics into axiomatics, that is, a static structure. Hilbert's views on geometry encouraged a consolidation of it as what does not have an essential connection to movement (as a line can also be named a chair). Fulfilling Hilbert's program for an axiomatically consistent geometry, Tarski had come to speak of geometry in mechanical terms. Tarski no longer refers to geometry as the study of space (together with constructions in and through it); he rather refers to its meta-properties as a static structure. Following Fuller, one may

---

26 See Hilbert 1899, section 9.
27 Tarski 1967, 306.
28 Tarski/Givant 1999, 184, axiom 10$_3$.
29 Heath 1908a, 155.
30 Note that in *Grundlagen der Geometrie* a line is a collection of points but also functions as a basic object.
31 Cf. Hilbert's reference to Kant's citation regarding the origin of abstract ideas from intuition (Hilbert 1899, 1).

say the Greeks' static constructs (e.g. the square or the cube) were replaced by a static structure for geometry itself.

2.3 The two sides of the mathematization of the fold at the 19th century

In light of a static conception of geometry, we ask how folding, as a dynamic operation, was perceived mathematically starting from the 19th century. Before turning to Fuller's conception of the fold, we will shortly examine the dual role folding played in mathematics at that time. This will help us situate Fuller's thought within a pertinent historical tradition.

A folded piece (of paper, fabric etc.) is regarded as such when one or two of the following operations are involved: creasing (as in folding a paper by a mountain- or valley-fold) or bending (without introducing creases). In this section we provide two examples of 19th century mathematizations of folding that took both operations into account: Sundara Row in his 1893 manuscript *Geometrical exercises in paper folding*, and Leonhard Euler, who described developable surfaces as folded. These mathematizations considered folding not only as a mathematical tool, but also as what expresses essential characters of the geometric form.

2.3.1 Row's Folds and the emergence of the physical straight line
Tandalam Sundara Row was an Indian mathematician, who worked for the Indian government in the revenue department. Row is mainly known for his book *Geometrical exercises in paper folding*.[32] Klein's favorable mention of Row's work in *Vorlesungen über ausgewählte Fragen der Elementargeometrie* sparked a general interest in the geometry of paper folding.[33] Why was Klein so impressed by Row's work on folding? To answer this question, let us examine how Row deals with geometry. To begin with, Row refers to the folding of paper as "kindergarten gifts" (the word 'Origami' does not feature). He invokes Friedrich Fröbel's gifts and occupations: "[t]he idea of this book was suggested to me by Kindergarten Gift No. VIII. Paper-folding".[34] Row states that "[t]hese exercises do not require mathematical instruments," referring to the straightedge and compass used in Euclidean geometry.[35] Row also dispenses with the need for axioms:

> "The teaching of plane geometry in schools can be made very inter-
> esting by the free use of the kindergarten gifts .... [the paper folding]

---

32 Cf. Friedman 2016 for a detailed account of Row's life and work.

33 Klein 1897, 42: "[...] we may mention a new and very simple method of effecting certain constructions, paper folding. [...] Sundara Row, of Madras, published a little book Geometrical Exercises in Paper Folding [...] , in which the same idea is considerably developed."

34 Row 1893, vii.

35 Ibid.

would give them [school children] neat and accurate figures, and impress the truth of the propositions forcibly on their minds. It would not be necessary to take any statement on trust." [36]

Row suggests teaching Euclidean geometry to children could be done without axioms, that is, without "statement[s] [taken] on trust." In comparison to Euclid, Row proposes a different conception of geometry: a geometry not grounded in axioms or ideal objects, but rather based on folding as its one and only allowable operation. As a result, the status of the straight line, as a product and producer at the same time, becomes clearer.

The opening chapter to Row's manuscript starts with a description of materiality, not with any foundational system of axioms:

> "Look at the irregularly shaped piece of paper [...] and at this page which
> is rectangular. Let us try and shape the former paper like the latter.
> Place the irregularly shaped piece of paper upon the table, and fold
> it flat upon itself. Let X'X be the crease thus formed. *It is straight.*" [37]

Row starts with an operation based on paper and hence on materiality: the folding of an "irregularly shaped" sheet of paper and later the passing of a knife. [38] The important point to consider here is that the line produced is straight as a direct result of folding. [39] There is no need to prove the line is straight, or define it as what passes through two points.

As the line X'X is only considered a consequence of folding, it obtains another status: it is that along which we fold: "Fold the paper again as before along BY, so that the edge X'X is doubled upon itself." [40] Row now folds the paper along the line that was just created, such that a part of this line X'X will be folded upon itself. When considering the crease BY that is created, Row discovers that BY and X'X are perpendicular. [41] Creating thus a rectangle, Row continues with the folding of a square whose side is of unit length. Then a smaller square is folded inside, rotated by 45 degrees in relation to the initial square. The process continues repetitively, creating via folding a sequence of squares embedded one into the other.

In Row's treatment, the straight physical line acquires a special status: it is at once created by the fold and creating it. It is crucial to emphasize Row always deals with line

---

36 Ibid, viii.
37 Ibid, 1 (our italics).
38 Ibid.
39 In contrast to Kempe's 1887 treatment of straight lines (Kempe 1887, 2–3).
40 Row 1893, 1.
41 "Unfolding the paper, we see that the crease BY is at right angles to the edge X'X." (ibid).

segments – the inevitable result of folding a piece of paper of finite dimensions. There are neither infinite lines (and hence no dispute over the parallel axiom), nor basic objects to begin with. There is rather a basic operation that initiates geometry. The basic objects, the *Grundbegriffe* and the relations between them do not play the same crucial role in Row's book, as they did for many of his contemporaries. Row takes into account neither group theory nor axiomatic methods he was surely well aware of.[42] In Row's work it is the fold – as what causes the *discrete*, finite, straight line to emerge as a material, *discrete* unit – that plays the crucial role.

2.3.2 Euler, folded surfaces and differential geometry

Let us now turn to developable surfaces: in this context the fold is considered a continuous operation. The history of developable surfaces can be traced as far back as Aristotle (384–322 B.C.).[43] In their current definition, developable surfaces are regarded as a special type of ruled surfaces: they have zero Gaussian curvature and can be mapped onto the plane without distorting curves.[44] Though the history of developable surfaces deserves a detailed account, we will only provide a brief survey focusing on their relation to folding.[45] In his development of calculus, Leonhard Euler (1707–1783) initiated the first serious mathematical study of ruled surfaces. He wrote his celebrated manuscript *About solids, the surfaces of which can be developed on the plane* – in the original: *De solidis quorum superficiem in planum explicare licet* – where he identified surfaces as boundaries of solids. Euler opened the manuscript with the statement that cylinders and cones have the property that they can be flattened out or "developed on the plane" unlike spheres. Euler wished to know which other surfaces share this property.[46]

It is important to note for the purpose of our discussion that *explicare* in Latin means 'to explain', 'to develop' but also 'to unfold'. The expression "in planum explicare," which features all throughout the paper,[47] can be translated verbatim into 'to unfold onto a plane'. The term 'developable surfaces' is a later nomenclature.

Euler failed to find developable surfaces (besides cylinders and cones) through analytical means. Using geometric principles, however, he did reach a solution. Employing geometrical results, Euler understands that lines that were parallel on the flat paper will also not meet on the folded one, concluding that the line element of the surface has

---

42 Row's awareness of other mathematical methods can be seen in Row 1906. Note the same year (1893) another manuscript on folding was published by the mathematician Hermann Wiener. See: Friedman 2016.

43 Aristotle states in *De Anima* that "a line by its motion produces a surface" (Aristotle 1928–1952, vol. 3, 409a).

44 Gaussian curvature is defined as the product of the two principal curvatures, which are the eigenvalues of the second fundamental form of the surface in question (the second fundamental form being a quadratic form defined on the tangent plane to a point on the surface). See e.g. Pressely 2001, 147.

45 For more detailed surveys see: Cajori 1929; Reich 2007; Lawrence 2011.

46 In Euler's words, "quorum superficiem itidem in planum explicare licet." In: Euler 1772, 3.

47 Ibid, 7, 8, 11, 27, 31 and 34.

to be the same as the line element of the plane. What is surprising perhaps is that the geometric principles in question were inspired by folded paper: "charta plicae".[48] It was folded paper and not solids, which informed the intuition behind developable surfaces in their early incarnation.[49]

Euler was not the only one to employ such terminology: the mathematician Gaspard Monge (1746–1818) also studied developable surfaces at the time, and, as with the former, described developable surface (and curves on them) as *pliée*, i.e., 'folded'.[50] It might be claimed that mathematicians (e.g. Monge, Euler) considered folding during those decades an essential action for creating surfaces, as an operation grounded in the materiality of the paper. However, it is important to remark that, with the further development of calculus and the rise of differential geometry, the term *Manifold* (*Mannigfaltigkeit*), albeit having an etymological connection to 'fold', was not chosen to describe surfaces as inherently *folded*. In his 1854 talk *Über die Hypothesen, welche der Geometrie zu Grunde liegen*,[51] Bernhard Riemann used the term *Mannigfaltigkeit* almost synonymously with 'magnitude', when he stated he set himself "the task of constructing the notion of a multiply extended magnitude,"[52] and invoked various motivations when first using the term. 'Mannigfaltigkeit' for Riemann can equally be discrete; it does not necessarily refer to a surface. When talking about continuous manifolds, the intuitions Riemann provides for choosing the term "Mannigfaltigkeit" are positions of objects and colors. No wonder a developable surface was and is considered a manifold and not a folded piece of paper.

3. Fuller's mobile structures

As was seen in sections 2.2 and 2.3, a withdrawal from materiality occurred in geometry at the end of the 19th century: consider for example Tarski's obvious mechanization of geometry. Row's manuscript on the other hand was either completely ignored or criticized for being "too infantile for a grown person."[53] Against this background, Fuller suggested that stable geometry (in the form of planes and lines) emerges in fact from mobile moving folds, threads and transformations.

---

48 Ibid, 7.
49 Euler was of course also one of the founding fathers of topology, along Henri Poincaré , Solomon Lefschetz and Johann Listing. Fuller was interested in topological transformations (e.g. the Jitterbug transformation, see section 3.5) and was aware of Euler's polyhedron formula: $V - E + F = 2$ (see section 3.4).
50 For example in: *Mémoire sur les développées, les rayons de courbure, et les différens genres d'inflexions des courbes a double courbure* (Monge 1785, 517–519); *Application de l'analyse a la géométrie, a l'usage de l'Ecole impériale polytechnique* (Monge 1809, 348 among others), *Géographie descriptive* (Monge 1811, 141).
51 Riemann 1868. Cf. also Cantor 1878, where it can be said that both mathematicians took manifolds as sets.
52 Riemann 1868, 133: "Ich habe mir daher zunächst die Aufgabe gestellt, den Begriff einer mehrfach ausgedehnten Größe aus allgemeinen Größenbegriffen zu construiren."
53 Young/Young 1905, vii.

3.1 Fuller and Semper: folds and interlaces

Folds and folding are not the primary consideration in Fuller's work. However, the most characteristic of his artifacts – should they be experimental buildings, maps or geometric modeling – are indeed folded or otherwise rely on folding as a deforming operation, as can be seen in fig. 1.[54] Considering how vitally important folds and folding were for Fuller's practical design, his remarks on the issue were dispensed sparingly, with most dedicated to specific problems of folding, such as the great circles.[55] Where one would otherwise expect a theory of folding to accompany Fuller's rich discourse on design, it is only found implicitly in his artifacts and the geometry of the Synergetics.[56] This disproportionality calls upon us to rediscover a tacit theoretical foundation from which to reconstruct the fold and the deforming operation.

Fig. 1: Necklace-Dome: One of the first folded geodesic domes of Fuller, done in 1950.

---

54 See also fig. 6 and 7: the *Jitterbug transformation*.

55 Fuller 1975a (sections 450 – 9) demonstrates eight models (a cuboctahedron and an octahedron, among others) that can be constructed by folding whole circles (with a protractor, using origami-style folding). See Fuller 1975a, section 459.03: "The six great circles of the icosahedron can be folded from central angles of 36 degrees each to form six pentagonal bow ties." Cf. also Fearnley 2009.

56 The changing and developing relationship between theory and design in Fuller's work is seen in: Krausse/Lichtenstein 1999; Krausse/Lichtenstein 2001.

In his earlier studies Fuller examined ways to reduce weight loads in architecture and construction. He was famous for provoking fellow architects with the question: "Does anybody know what a given building weighs?"[57] Weight load reduction, employed as a design strategy, was for him a means for examining economical and efficient construction. Fuller noted how weight specifications come up naturally in the design of marine vessels, vehicles and aircrafts, while in building construction this information is considered irrelevant. His question attempted to bridge the gap between the two practices (or mentalities): the mobile and the stationary.

Techniques for consolidation, folding and size or shape adaptation are present in all mobile forms of human habitation (such as tents, yurts and tipis). This is true not only for the architectural structures themselves, but also for equipment and furniture that go along with them. For a nomadic way of life, weight is not the only crucial criterion. Objects belonging to the household must fit requirements for transport. Folding fulfills these requirements in great measure: it allows objects to assume various shapes, being either flattened or spatially expanded. Folding allows for a transformation, with which objects can be adapted to mobility. Folds are thus both a result and an expression of movements, whose event-patterns Fuller summarizes under the concept of precession.[58]

One can observe firsthand the direct link between movement and folding in everyday clothes and textiles: dresses, cloaks, curtains, carpets and so on, as well as adjustable flexible space partitions.[59] The fact that, under this aspect of regulation between inner and outer, a systemic correspondence between organisms and artifacts can be devised, is not least suggested by the fact that the use of hides of animals and barks of trees belongs to one of the oldest techniques for space subdivision.

The architect Gottfried Semper (1803–1879), regarded as one of the originators of research into material culture, derived his theory of architecture from primitive artifacts, such as clothing, used for space partition. This theory finds expression in his monumental work *Style in the Technical and Tectonic Arts; or Practical Aesthetics* (1860–3).[60] What is of special

---

57   "Does anybody know what a given building weighs? I once asked an American Symposium of architects including Raymond Hood and Frank Lloyd Wright as well as the architects of Rockefeller Centre, the Empire State Building and the Chrysler Building what the different structures they were designing weighed. Clearly, weight was not one of their considerations. They didn't know." In: Fuller 1963, 53.

58   "The effects of all components of Universe in motion upon any other component in motion is precession, and in as much as all the component patterns of Universe seem to be motion patterns, is whatever degree they affect one another, they are inter-affecting one another precessionaly, and they are bringing about resultants other than 180 degrees. Precess means that two or more bodies move in an interrelationship pattern of other than 180 degrees." In: Fuller 1975a (section 533.01), 287.

59   Fuller encompasses an overarching notion of a dwelling place with "environment controls." See Fuller 1963, 55 ff; Compare Krausse 2002b, 97 ff.

60   Semper 2004.

interest for us here is his account of the textile origins of architectural space enclosures, exemplified through the wall as an architectural element. In his draft of 1853 he writes:

> "We have in our German language a word which signifies the visible part of the wall, we call this part of the wall, *die Wand*, a word which as a common root and is nearly the same with *Gewand* which signifies woven stuff; the constructive part of the wall has another name, we call it *Mauer*. This is very denoting."[61]

"Denoting" hence distinctive. Here one finds not only two classes of materials – fabric and fiber on the one hand, rocks and soil on the other – but also two different principles of structure, reflected in two different types of construction. While hard crystalline materials tend to resist compressive forces till they give way to pressure in the form of fractures and fissures, fiber-based materials absorb tensile, attractive forces and bending stresses; in contrast to crystalline materials they are flexible. Semper shows in his early writings, it is the latter that preceded masonry.

> "It is a fact," he writes, "that the first attempts of industrial art, which have been made and which we still observe to be made by human beings, standing on the sill of civilization, are *dresses* and *mats*. This part of industry is observed to be known even by tribes, which have no idea of *dressing*."[62]

Plaits, carpets, interlaces and hangings were originally used for space arrangement and partition, to which solid structures were subsequently added,

> "the thick stone-walls, were only necessary with respect to other secondary considerations, as for instance to give strength, stability, security etc. Where these secondary considerations had no place, there remained the hangings the only means of separation; and even when the first became necessary, they formed only the inner scaffold of the true representative of the walls, namely the variegated hangings and tapestries."[63]

Semper demonstrated how these elements enable flexible interior partitions.[64] Flexibility and mobility originally form a unit that is lost with the use of solid structures, and must

---

61 Semper 1983, 21.
62 Ibid.
63 Ibid.
64 As an example, Semper cites the Caribbean hut in which the walls are transferable and not connected to the roof. Cf. Semper 1986, 34f.

be compensated for using doors. This idea experienced a modernist revival in the form of mobile room dividers and separators, sliding doors and accordion folding partitions. A most striking example is found in *curtain walls*, whose construction in recessed reinforced concrete columns (as in Bauhaus Dessau 1926) accounts for the old truth: the space-enclosing elements of architecture are in effect suspension structures subject to tension.

3.2 Fuller's non-simultaneous foldings

Semper's reintroduction of fiber-based materials into the processing and manipulation of form was taken on by Fuller; this time of course under the conditions of advanced industrialization, new materials, innovative construction techniques and global transportation systems. Fuller defines the fundamental relationship of human existence to mobility as follows:

> "Man was designed with legs – not roots. He is destined to ever-increasing freedom of individually selected motions, articulated in preferred directions, as his spaceship, *Earth*, spinning its equator at 1000 miles per hour, orbits the sun at one million miles per day, as all the while the quadrillions of atomic components of which man is composed inter-gyrate and transform at seven million miles per hour. Both man and universe are indeed complex aggregated of motion."[65]

This is a concise summary of what Fuller called *scenario universe*. It is this scenario that forms an indispensable part of the exposition to Fuller's energetic-synergetic geometry.[66] A scenario is favored over theorems or axioms; it emphasizes the a priori temporality of a (geometrical) event:

> "The Universe", so presented in his book *Synergetics*, "can only be thought of competently in terms of a great unending, but finite scenario whose as yet unfilled film-strip is constantly self-regenerative [...]. Our Universe is finite but non-simultaneously conceptual: a moving picture scenario of non-simultaneous and only partially overlapping events".[67]

The reference to the scenario and to the agility and mobility of the film expresses Fuller's deep mistrust in the image, the still image, the single frame with its implied immobility.

---

65 Fuller 1969, 348.
66 Fuller 1975a (section 320.01–02), 87.
67 Ibid.

The single image evokes the illusion of simultaneity of events, as in the image of the starry sky, an image that exhibits light, which in fact emanated from stars at different moments in time.

Only when considering longer time spans can one observe evolution and metamorphosis in nature. An emphasis on non-simultaneity (with partial overlapping at the most) is also present in the way 'folding' is present in the German words *Überlegung* and *überlegen*. The verb *überlegen* has indeed the following three different meanings: 1) cover; coat; overlap, 2) consider; contemplate, 3) lay an object down over another object.[68]

The introduction of the scenario, as a form of thinking and of presentation, allows us to work with partially overlapping events, where scenarios are both descriptive and prescriptive – prescriptive with respect to actions that would be performed, carried out and so executed.[69] The performative aspect of the scenario also plays a role in Fuller's geometry, which insists on the embodiment and the materialization of geometric figures in the model, as well as in the live-performance of transformations that he discovered. How did Fuller come to adopt the scenario as a framework for cognition? We already detect its origins in his first architectural project, as a framework for design. The structure, which he has in mind, is not developed in response to the layout of its designated lot, but in accordance with easy transport. The tower house, which was designed in 1928, could be industrially prefabricated and then shipped by air (with a Zeppelin); it could be delivered to any location on the globe. Fuller, even before clarifying what was needed and implied in constructing such a house, first simulated this unprecedented procedure. To that end, Fuller drew a series of sketches in the manner of a comic strip, which depicted the key events of this scenario.[70]

Fuller's recordings from this period show how attentively he followed the development of this popular genre and reflects on its potential as a form of presentation: "Undeniably the 'funnies' are the most generally inspected portions of our daily newspapers and may be considered the economic frosting that sells the cake – It is more than significant that these funnies have completely lost race of 'slapstick' and have become serials of handy philosophy."[71] Even later on, in his preparation of maps and diagrams of complex global data (world energy map, global transport development, history of isolation of chemical elements), Fuller insisted on "maintaining a comic strip lucidity".[72]

---

68 Grimm 1936, column 385.
69 Regarding the various aspects of performance in Fuller's work, cf. Krausse 2016.
70 Krausse/Lichtenstein 1999, 99–103.
71 Krausse/Lichtenstein 2001, 102.
72 Ibid, 152 (from Fuller's *Earth incorporated* (1947)).

Traces of scenario-thinking, partial overlapping and comic-strip lucidity can be found also in Fuller's pictorial depiction of the construciton process. Take for example the sequence of photos that illustrated the construction of the *Dymaxion house* starting from its components, through the individual assembling steps, to the finished and furnished residential house. The same pattern appeared in his second attempt to build this house – albeit with an altered outline – in an aircraft factory. This time it was indeed realised as a prototype.[73] Besides the process of assembling, the most important thing about this sequence of photos is its demonstration of initial and final construction stages: the initial assembly of lightweight thin parts, occupying but little space, set against the finished space-consuming building. Before construction commences, building parts are laid out as one spreads clothes before packing a suitcase.

In this way one may inspect all components in order; they were designed to fit into a container in the most space efficient way. In the case of the *Wichita house* of 1946, the cylindrical, metal, storage container served also a key structural element of the building. Packaging aligns well with the concept and practice of folding. Transportation to and unpacking at the construction site need to be taken into account in the design of the container and its contents. Unpacking should fit color-coded step-by-step assembly all the way up to the finished, fully furnished, turn-key house. This turns building into a performance that follows a precise scenario.[74]

It is no coincidence, that this performance, as in a sequence of movie frames, resembles the process of the unfolding of a plant from seed to bud to leaf, save that its origin goes back to design, from which mechanical parts are developed as affiliated and connected.

3.3 Seedpods, Viruses and Geodesic domes

Fuller related design scenarios to organic growth processes on various occasions. One of his experimental constructions, the *Flying Seedpod* of 1953, is a pure folding mechanism.[75] Flying Seedpod is a dome, 42 feet in diameter, which can set itself up semi-automatically. Whether compacted as a transportable bundle or deployed as an architectural structure,

---

73 The corresponding photo series of the Dymaxion House and the Wichita dwelling machine are printed in: Marks 1960, 84 f and 128–133.

74 Fuller's scenario-thinking looks beyond the finished product onto its ultimate end-of-use. With its structures he envisages "demountability", with its materials, "recirculation." Responsibility in design extends to the entire life cycle of the product – what Fuller called "cradle-to-grave." It took the combination of product-cycle together with recycling to go from "cradle-to-grave" to the slogan "cradle-to-cradle". Regarding "demountability" cf. Marks 1960, 112–113. Regarding recycling, cf. Fuller 1938, chapter 38, 316–322; Reprint in: Krausse/Lichtenstein 2001, 117–120. Regarding "cradle-to-cradle", cf. Braungart/McDonough 2009.

75 See fig. 2: series of four photos "Flying Seedpod" 1954/5.

all parts stay connected to each other by flexible nodes or joints. The bundle consists of 30 inwardly folded tripods, whose chains come together in ball joints. The system of tripods can be unfolded and straightened up by extending pistons in pneumatic cylinders – radially directed tubes at the vertices of the dome. With the extension of pistons, a net made of cables is stretched. The clear span unsupported dome-structure obtains firmness and rigidity, through the interaction of its push-pull components. Flying Seedpod was a project that Fuller realized in 1953 with students from Washington University (fig. 2).

Fig. 2: Flying Seedpod. Washington University, St. Louis, 1953; A folding-out geodesic structure.

The study of folding structures of geodesic domes developed alongside progress in space exploration missions, so that one might see in Flying Seedpods – "the first scientifically designed apartment" – a rocket capsule to the moon.[76]

Though the seedpod was nicknamed "the moon structure," it did not fly to the moon. Instead it appeared in other ways in the world of molecular biology. Fuller tells how it came about:

---

76 "You may possibly be looking at the prototype of the structural principles that we may use in sending history's first (little) scientific dwelling to the moon. As you see, all the structural members are tightly bundles together in parallel so that they may be transported in minimum volume within a rocket capsule." In: Fuller 1965, 70; Fuller's foundations for folding structures were later continued by his pupil Joe Clinton for NASA. Cf. Clinton 1971.

"The principle of structural dynamics of the [...] moon structure, the flying seedpod and its logistic pattern transformability, are double interesting because they have turned out to be also the same structural, self-realization system employed by a class of microcosmic structures – the protein shells of all the different types of viruses. About three and one half years ago molecular biologists in England and their colleagues in America, working in teams, were trying to discover the structural characteristics of the protein shells of the viruses with X-ray diffraction photographic analysis. These virus scientists discovered that the viruses' protein shells were all some type of spherical geodesic structure. Having previously seen published pictures of my geodesic structures they corresponded with me and I was able to give them the mathematics and show them how and why these structures occur and behave as they do. They have now found the poliovirus structure to be the same structure as the possible 'moon structure'. The polio virus instead of having the tripods on the outside and the clusters of five and six feet on the inside, has the five – and six – way jointings outside and the tripods or three–ways on the inside."[77]

The encounter between Fuller's experimental architectural structures and science of the day could be passed for incidental – a random correspondence between structures on widely disparate scales – if not for geometry that provides a connection of a more general nature.

The researchers at Birkbeck College in London engaged with a striking resemblance between viral capsids and Fuller's geodesic domes,[78] when the largest was just completed, almost 120 m in diameter, making it the largest ever built clear span dome.[79] It appeared as if the same such spherical structure found in nature was anticipated by Fuller, or rather as if he had built his geodesic domes according to models from nature. The first recorded images of capsids produced by an electron microscope were published in 1962; they finally made resemblance evident.[80] On this basis and other of Fuller's 1960

---

77 Fuller 1965, 72.
78 Cf. Morgan 2003, 86: "In the mid 1950s, Francis Crick and James Watson attempted to explain the structure of spherical viruses. [...] biophysical and electron micrographic data suggested that many viruses had > 60 subunits. Drawing inspiration from [Fuller's geodesic domes and] architecture, Donald Caspar and Aaron Klug [...] proposed that spherical viruses were structured like miniature geodesic domes," by forming a (protein) shell. Indeed, "[t]he idea was that identical viral subunits could bind together in quasi-equivalent positions to form a shell with > 60 subunits while conserving the same specific contact pattern between subunits" (ibid, 88).
79 Union Tankcar Company Geodesicdomes, Baton Rouge, Louisiana; Railway repair facility, October 1958, cf. Marks 1960, 222f.
80 Ubell 1962, 1.

architectural structures, the mathematician Harold Scott MacDonald Coxeter was able to match individual domes, each slightly different in its geometric resolution of geodetic networks, to individual types of viruses, whose capsids likewise vary geometrically.[81] This new visual input, brought to bear through advances in electron microscopy (EM), made the resemblance even more apparent in microorganisms. Fuller got to see electron microscope images of marine microorganisms (magnified up to about 50,000 times), taken by biologist Gerhard Helmcke (who specialized in lightweight constructions in nature), in a 1962 meeting together with his colleague Frei Otto and Helmcke himself. The architect Frei Otto, who had just launched the research group *Biology and Building*, later reported how impressed Fuller was:

> "The stereoscopic photographs looked like models of [Fuller's] famous domes. To the participants it was clear: Had he [Fuller] known the diatom shells before, the whole world would have said that he had learnt this by watching the living nature. Had he knew the diatom shells, how they were really, he would not have probably dared to build his shells."[82]

When Fuller met Aaron Klug and his interdisciplinary team of researchers in July 1959 there were no clear images yet, only clues coming from the X-Ray analysis of crystals. Even when Klug succeeded in applying crystallographic EM to the analysis of complex viral capsids, the images obtained were rather confusing: due to its extensive depth of focus, all structures were depicted one on top the other. Klug, who received the Nobel Prize in Chemistry in 1982 for this work and others, remarked in his Nobel lecture: "Thus, we knew what we were looking for, but we soon found that we did not understand what we were looking at".[83] Fuller's geodesic domes served not only as a possible guide to deciphering those EM images. They were geometrical models at large which allowed patter recognition of an unidentified micro-phenomenon.

3.4 Platonic solids: a stable habitat?

The point of contact between Fuller's designs and structural chemistry is derived from a functional correspondence: both deal with a problem of habitation that needs to be solved with utmost care for resources – one might consider protein shells to be the smallest houses in nature. In the case of the virus the space within is occupied by DNA and RNA molecules, which are densely folded and packed waiting for a suitable host to open the

---

81   Coxeter 1971. On the relationship between Coxeter and Fuller, see Roberts 2006, chapter 9.
82   Otto 1985, 8.
83   Klug 1992, 89.

protein shell. Nevertheless, both Fuller's domes and capsids represent an attempt to solve an optimization problem – maximum capacity for the smallest surface area. The geometric solution to the problem is the sphere, which both Fuller and viruses opt for.[84] Turning now to the structural elements of an approximately spherical casing or shell, a method of regular subdivision of the sphere must be developed.

In subdividing the sphere, Fuller – similarly to viruses, marine microorganisms and carbon molecules – chooses another way, compared with dividing the globe into a network of longitudes and latitudes. He avoids the classical construction of a dome using meridian groins and horizontal rings or bands. Instead, the sphere is symmetrically divided into regular polygons, described by Plato as the Elements in his dialog *Timaeus*.

Common to all Platonic solids – the tetrahedron, hexahedron (cube), octahedron, dodecahedron and icosahedron – is that their vertices lie on a circumscribed sphere and their edges, once projected onto the circumscribed sphere, form arc segments of equal length. The arc segments are all part of great circles. Roughly speaking, great circles are the paths of minimal length on the sphere. They correspond to the straight lines of Euclidean geometry. Together, great circles and lines belong to a class of paths known as geodesics. Back to Platonic solids, the system of geodesic segments, obtained through projection onto the circumscribed sphere, forms a regular grid that divides the surface of the ball into equal polygons.

With the icosahedron, one obtains the most tightly arranged subdivision; it consists of 20 equilateral triangles adjoining along 30 edges and touching at 12 vertices. According to Euler's characteristic formula, $V - E + F = 2$, where V denotes the number of vertices, E the number and F the number of faces. Five edges meet at the vertices of the icosahedron. A fivefold rotation symmetry is maintained throughout all of its subdivisions. This becomes clear when one looks at the truncated icosahedron:[85] the 12 vertices are trimmed; one third of each edge is truncated at each of both ends, resulting in a new polyhedron consisting of 12 pentagons and 20 hexagons, with each pentagon surrounded by five regular hexagons. The truncation of the icosahedron affords a way to refine the spherical subdivision thereby better approximating a sphere. The truncated icosahedron is now well known for the Telstar football and the discovery of the carbon molecule C_{60}. The 1996 chemistry Nobel laureates, Harold Kroto, Richard Smalley and Robert Curl, responsible for the discovery, named it *Buckminsterfullerene* in recognition of the architect's work.[86]

---

84 We leave aside a class of rod-shaped viruses such as the prototypical tobacco mosaic virus.
85 Comprising 32 faces, 90 edges and 60 vertices.
86 Kroto 1996. See also Krausse 2002a.

3.5 From folding and foldable platonic solids to the Jitterbug transformation

A question that occupied virologists was whether viral capsids – comprising at least 120 protein subunits under seemingly identical conditions – contrived their structure in a manner similar to Fuller's geodesic domes, domes which might have more than 180 of triangular subunits. These domes are able to modify themselves slightly, such that they arrange themselves according to a geodesic grid on a sphere.[87] Fuller could clarify this through his grid and through a formula, describing the total number of edges of these triangulated domes. Situating at every corner of a triangle a ball of radius 1, Fuller imagined a growing shell structure, beginning with the 12 balls, whose center situated on the surface of a sphere, then while the structure grows and another layer is situated symmetrically on the outside of the former layer, there are 42 balls, then 92 balls and so on, according to the formula: $n = 10f^2 + 2$. With this formula, Fuller bases his calculation on the *cuboctahedron*, being one of the 14 semi-regular Archimedean polyhedra: while inscribing a cuboctahedron with edge length 1 in a sphere, Fuller instructs, as explained above, to posit 12 balls centered on the vertices. As one may enlarge the sphere and the cuboctahedron (when now the edge length of the cuboctahedron is 2), one may posit 42 balls centered on the vertices, edges and faces of the cuboctahedron. In Fuller's formula n stands for the balls situated symmetrically in the growing shells, where f stands for the number of the layers of these growing structures.[88] And the higher the frequency – that is, the number of layers, the greater the number of balls and hence triangular subunits (created by drawing a line between three adjacent, tangent balls) is and the more fine-meshed the network of geodetic structures. Fuller adds that "[t]hese successive layers, which permeate each other in all directions may be identified with energy waves radiant in all directions from a nucleus."[89] By "nucleus" Fuller points to the fact, that while situating the 12 balls of radius 1 on the sphere, having their center as the vertices of the cuboctahedron, there is a room for an additional ball – called "nucleus", located exactly in the center of the sphere and touches all the other 12 balls (see fig. 5).

Fuller's mathematical model was therefore not developed from the icosahedron or from any of the other Platonic solids: as we will explain later, when balls are positioned at the vertices of the icosahedron, the structure thus obtained does not have a nucleus, from which "energy waves" emanate. To see how a nucleus is necessary, one constructs

---

87 "[...] we have discovered that the way these viruses were built was similar to the way the geodesic domes were Built. Geometrically you cannot put more than 60 identical units on the surface of a sphere with each one making identical contacts [...] The virus we had been working on had 180 sub-units – three times 60 – so they couldn't all be in identical environments." In: Klug 1995, 10.

88 For a detailed explanation, see: Edmondson 1987, 114–116. For a proof of Fuller's statement: Coxeter 1974.

89 Krausse/Lichtenstein 2001, 169.

dense *sphere packing* using identical spheres. By connecting the centers of spheres in certain arrangements one can obtain the elementary forms, for example Platonic solids.

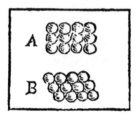

Fig. 3: Kepler's two types of planar packing of spheres.

Sphere-packing arrangements – being an arrangement of non-overlapping, possibly touching spheres – go back to Johannes Kepler's 1611 book *De Nive sexangula*. In it Kepler distinguishes between two types of packing: one that forms a cube and another that forms a tetrahedron.[90] Kepler's book is primarily an investigation into the hexagonal form of snowflakes. As to the densest space-filling arrangement of spheres, Kepler conjectured that the tightest packing produces rhomboidal aggregates, known as the *rhombic dodecahedral honeycomb*. The density η of a packing of solid spheres is today defined as:

$$\eta = \lim_{t \to \infty} \frac{\sum_{i=1}^{\infty} \mu\,(K_i \cap B_t)}{\mu\,(B_t)}$$

where $\mu(X)$ is the volume of X, B_t is a ball of radius t centered at the origin, and K_i are balls which are used for the packing.[91] It follows that the density of a packing of balls is always smaller then 1. When Kepler discusses sphere packing, he proposes two types. The first is the simple cubic packing and the second is what is called today an *FCC packing*, i.e. the hexagonal arrangement. On the cubic arrangement, Kepler concludes: "The arrangement will be cubic, and the pellets, when subjected to pressure, will become cubes. But this will not be the tightest pack." However, when considering the second packing, Kepler remarks that "[t]his arrangement will be more comparable to the octahedron and pyramid. The packing will be the tightest possible, so that in no other arrangement could more pellets be stuffed into the same container".[92] It is known today that the density of the FCC packing is $\pi/(3\sqrt{2}) \sim 0.7405$, whereas the density of the simple cubic

---

90   See: Kepler 1966. See also fig. 3.
91   See e.g. Conway/Sloane 2013, 8.
92   Kepler [1611] 1966, 15.

packing is approximately 0.6802, but Kepler does not give any reason why this pyramidal arrangement is the densest. [93]

Fuller studied sphere packing through his work on the suspended construction of the Dymaxion House. For this purpose, he used cables wrapped around an inner rope. When cross-sectioning the cables, one sees that the cross-section consists accordingly of six units around one center ("six around one". See fig. 4b).[94] In this cross-sectioning Fuller sees a prototype of symmetrical growth. Fuller's first study of a wave-mechanical matrix appears in grids for his hexagonal layout of the house on a mast, in which the intervals are specified not only in length dimensions, but also in time units.[95] The initial intuition for the *matrix* (or *isotropic vector matrix*) is found in the hexagonal configuration packed with circles: it belongs to the class of two-dimensional densest packing. Connecting the centers of adjacent circles with lines, one obtains a part of a configuration of 9 triangles, which the Pythagoreans called *tetractys*, as can be seen in fig. 4a.

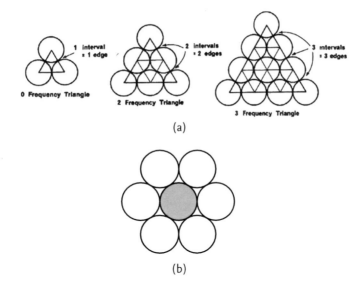

Fig. 4: (a) Only with the most right image (the *tetractys*) one may notice the appearance of the nucleus, in the middle of the inscribed hexagon. (b) Fuller's "six around one" construction, obtained from the tetractys. The nucleus is the grey circle.

---

93 Kepler's assertion, better known as the Kepler's conjecture (that the densest packing of identical balls in space is either a cubic arrangement or an hexagonal arrangement), was proved only in 2014 by Thomas Hales.

94 In the magazine he edited SHELTER (November 1932, 106–107) Fuller assembled pictures that, inter alia, show snow crystals, cable cross-sections and the Dymaxion House hexagonal plan. See facsimile reprint in: Krause/Lichtenstein 1999, 172–173.

95 Krause/Lichtenstein 1999, 114 f. "Time based plan for the 4D House", Figure 1928.

When extending this method to three-dimensional densest packing modeled on the cuboctahedron, Fuller indicates:

> "When the centers of equiradius spheres in closest packing are joined with lines [i.e. a possibly infinite truss modeled as the cuboctahedron], an *isotropic vector matrix* is formed. This constitutes an array of equilateral triangles which is seen as the comprehensive coordination frame of reference of nature's most economical, most comfortable structural interrelationships employing 60-degree association and disassociation." [96]

What Fuller searched for becomes clear in his 1938 book *Nine Chains to the Moon*. There Fuller calls for a time-based geometry that takes into account propagation of waves and rays and growth processes in space and time: "Time, or how far (or more properly 'fast') radially outward, in time and space, integrated as rate or the center of the sphere." [97] The mental image Fuller uses is that of the "expanding sphere" or the "halo" – a radiation in all directions. As was seen above, Fuller finds a framework for such processes in a polyhedron consisting of 8 triangles and 6 squares: the *cuboctahedron*. In its representation using sphere packing there are not "six around one" but rather "twelve around one." [98] Its shell consists of 12 balls. Growing symmetrically with additional layers, 42, 92, 162, 362 (or more) spheres can be packed. The cuboctahedron may be regarded a form of sphere packing: it forms a shell of 12 balls with a nucleus. Having a different shape compared with that of the cuboctahedron, one might say the icosahedron consists merely of a shell, whereas the cuboctahedron has its nucleus, as noted above and as can be seen in fig. 5.

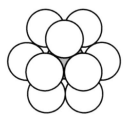

Fig. 5: On the left, a packing of 12 white balls, where the center of each (white) ball is placed on a vertex of a cuboctahedron; note the existence of a grey nucleus. On the right, a packing of 12 balls, where the center of each ball is placed on a vertex of the icosahedron; in this case there is no space for a nucleus of the same size. Note that the density of the icosahedral packing is approx. 0.6882, being lower than the density of the FCC packing.

---

96   Fuller 1975a, caption to figure 420.02 (our italics).
97   Fuller 1938, 134.
98   Fuller 1975a, 116–120, section 413.00.

Since the hollowed space in the center of the icosahedron has smaller dimensions compared with the ball in the nucleus of the cuboctahedron, the radii of the balls layering the icosahedron are somewhat shorter than the edges. Therefore, the icosahedron cannot provide a basis – a matrix, in Fuller's terminology – for symmetrical growth processes. It always remains the same perfect space division, through all of its modifications: be it a capsid, a shell or a geodesic dome. Conversely, the cuboctahedron provides a matrix for growth processes – what Fuller refers to by "isotropic vector matrix" – but is unsuitable as a building structure. The six square faces of the cuboctahedron lack stabilizing diagonals. Fuller has repeatedly demonstrated this effect, for example, in his Necklace Performance.[99] A square, according to Fuller, is not a structure; it is a temporary opening at the most. Only the triangle is a structure, as it is self-stabilizing. This feature appears only when one builds the cuboctahedron as a model and connects the rods with flexible joints.

These and many other aspects come to light in a geometric transformation Fuller named *Jitterbug* after a 1940 popular dance. Though there are different ways to dance the Jitterbug,[100] we will limit ourselves to one, the easiest, which is performed with a model. The model consists of 24 individual rods that may move while being connected to flexible nodes. Thus, various configurations can be produced via folding. It begins with the most extended configuration: the cuboctahedron. Though the cuboctahedron also has square faces, we focus on the behavior of the triangles during the transformations (see fig. 6 and 7). Its 14 faces can be seen clearly, and it is evident via touching that the cuboctahedron is not rigid. When slight pressure is applied to the model's upper triangle, the result is a left or right rotation of the remaining triangles. This draws the 6 squares in a diagonal direction, deforming them into rhombi. At this point the squares crease, forming an invisible, "silent" edge, while an inserted rod could have prevented this deformation. When the rods are inserted in the middle of the creased squares, the result is of an icosahedron: each folded rhombus supplies 2 triangles. A total of 12 new triangles plus the original 8 add up to the 20 triangles of the icosahedron. In the absence of intervention to stop the process of folding it goes on: adjacent edges of the cuboctahedron's original squares join in pairs. The result is an octahedron, whose 8 triangles are formed with double-tipped edges. The next two stages of the Jitterbug transformation are more complicated since they require further extension and folding of area, suffice it to say, this procedure produces a tetrahedron, with quadruple edges and culminates in a planar triangle with eightfold edges. Throughout the process of transformation the polyhedra emerge in a process of phase transitions.

---

99 Fuller 1975a, 317–319, sections 608.00–609.01.
100 Cf. "Five ways to dance the jitterbug" in Krausse/Lichtenstein 2001, 24–33.

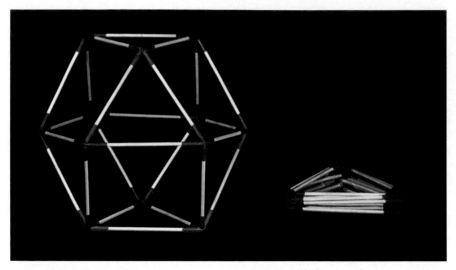

Fig. 6: The initial and the final position of the Jitterbug transformation cuboctahedron and triangle.

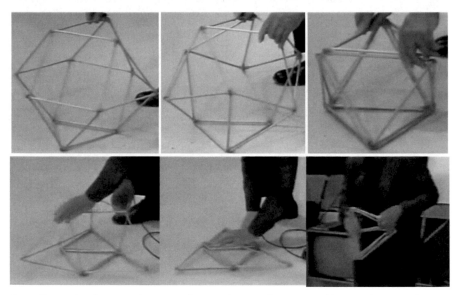

Fig. 7: Photos taken from R. Buckminster Fuller's 1975 lecture "The Vector Equilibrium", where several stages of the Jitterbug transformation are presented.

The Jitterbug transformation demonstrates convertibility in a rigid and stable structure; it demonstrates metamorphosis from one "solid" into the next in a single, continuous process of contraction and expansion up to the structure's limits. This metamorphosis not only offers a different view on geometry, it also offers another perspective on thought: equally abundant in forms, the Jitterbug transformation is also abundant in patterns of movement. Fuller demonstrates ongoing articulations of movement have a relationship to epistemology:

Michael Friedman, Joachim Krausse

"There is quite possible a scientific truth to be evolved from the fact that motion, particularly rhythmic motion, is highly provocative of thought objectivation. Certainly travel provides perspective, broad angles, and accelerated progression potential of clarification of the experience trend."[101]

4. Folding: The provocation of thought

Mobile structures, as with the Jitterbug transformation, are at the core of Fuller's thought. Opposing geometry's withdrawal from materiality and architecture's withdrawal from mobility, Fuller suggests a revival of both concepts back into geometry. Indeed, *partial* overlapping and non-simultaneity are pilled off via the axiomatized mechanized conception of geometry – as all lines and axioms appear at once – whereas the mathematization of the fold ignores its materiality as a guiding principle. Following Semper and taking the German word *Überlegung* as a cue, folding suggests an interlacing of thought and contemplation together with a materialized geometry and partially overlapping events. This enables Fuller to tie together his conception of geometry as a material mathematical science – a point is a place where two lines pass through but not at the same time – and the scenario as a form of thinking. Folding, weaving and interlacing engender this form of thinking. As Semper indicated, it is flexible, mobile, foldable interior partitions that enable the wall to be a pure structural element, not the other way around. Fuller takes on this viewpoint and pursues it into the macroscopic world (seedpods and the dance of human beings) as well as into the microscopic world (inspiring the discovery of the structure of viral capsids through his geodesic domes). It is indeed the same line of thought that is apparent in Fuller's work: geometry, whose essence in exemplified in the folding and unfolding of platonic solids as they metamorphose through the Jitterbug transformation, is not a rigid structure that lacks movement or consideration to materiality, but rather it is the thought-provoking *scenario universe* of material, flexible, non-simultaneous, partial overlapping.

---

101 Fuller 1938, 139.

Bibliography

Aristotle (1928–1952): *Complete Works*, vol. 1–12 Ed. and trans. by Ross, William David, Oxford: Oxford University Press.

Blumenthal, Otto (1935): *Lebensgeschichte*. In: Hilbert, David: Gesammelte Abhandlungen, Bd. 3: Analysis, Grundlagen der Mathematik, Physik, Verschiedenes, nebst einer Lebensgeschichte. Berlin: Springer, pp. 388–429.

Braungart, Michael / McDonough, William (2009): *Cradle to cradle. Remaking the way we make things*. London: Vintage Books.

Cajori, Florian (1929): *Generalisations in Geometry as Seen in the History of Developable Surfaces*. In: The American Mathematical Monthly, vol. 36, no. 8, pp. 431–437.

Cantor, Georg (1878): *Ein Beitrag zur Mannigfaltigkeitslehre*. In: Journal für reine und angewandte Mathematik (Crelles Journal), vol. 84, pp. 242–258.

Coxeter, Harold Scott MacDonald (1971): *Virus Macromolecules and Geodesic Domes*. In: Butler, John Charles (ed.): A Spectrum of Mathematics. Wellington: Auckland University Press, pp. 98–107.

Coxeter, Harold Scott MacDonald (1974): *Polyhedral Numbers*. In: Cohen, Robert S. / Stachel, John J. / Wartofsky, Marx W. (eds.): For Dirk Struik, Scientific, Historical and Political Essays in Honor of Dirk J. Struik (Boston Studies in the Philosophy of Science, vol. 15). Reidel: Dorrech, pp. 25–35.

Clinton, Joseph D. (1971): *Advanced Structural Geometry Studies. Part 2: A Geometric Transformation Concept for Expanding Rigid Structures*. Southern Ill. University, NASA Report CR-1735, Washington D.C..

Conway, John / Sloane, Neil J. A. (2013): *Sphere Packings, Lattices and Groups* (Grundlehren der mathematischen Wissenschaften, 290). New-York: Springer.

Corry, Leo (1997): *David Hilbert and the Axiomatization of Physics (1894–1905)*. In: Archive for History of Exact Sciences, vol. 51, pp. 81–97.

Corry, Leo (2004): *Modern Algebra and the Rise of Mathematical Strucutres*. 2nd ed. Basel: Birkhäuser.

De Risi, Vincenzo (ed.) (2015): *Mathematizing Space: The Objects of Geometry from Antiquity to the Early Modern Age*. Basel: Birkhäuser.

Edmondson, Amy C. (1987): *A Fuller Explanation: The Synergetic Geometry of R. Buckminster Fuller*. Boston: Birkhäuser.

Euler, Leonhard (1772): *De solidis quorum superficiem in planum explicare licet*. In: Novi commentarii academiae scientiarvm imperiatis Petropolitanae, vol. XVI, pp. 3–34.

Fearnley, Chris J. (2009): *Foldable Great Circle Geometries*. In: Hyperseeing, vol. 2, pp. 63–64.

Feferman, Solomon / Feferman Burdman, Anita (2004): *Alfred Tarski: Life and Work*. Cambridge: Cambridge University Press.

Friedman, Michael (2016): *Two beginnings of geometry and folding: Hermann Wiener and Sundara Row*. In: BSHM Bulletin: Journal of the British Society for the History of Mathematics. DOI: 10.1080 / 17498430.2015.1045700 (in print).

Fuller, R. Buckminster (1938): *Nine Chains to the Moon*. Philadelphia: Lippincott.

Fuller, R. Buckminster (1963): *The R.I.B.A. Discourse: Experimental probing of architectural initiative*. In: id.: Ideas and Integrities. Englewood Cliffs, N.J.: Prentice-Hall, pp. 35–66.

Fuller, R. Buckminster (1965): *Conceptuality of Fundamental Structures*. In: Gyorgy Kepes (ed.): Structure in Art and in Science. New York: George Braziller, pp. 66–80.

Fuller, R. Buckminster (1969): *Utopia or oblivion: the prospects for humanity*. New York: The Overlook Press.

Fuller, R. Buckminster (1975a): *Synergetics: Explorations in the Geometry of Thinking*. (In collaboration with E. J. Applewhite). New York: Macmillan.

Fuller, R. Buckminster (1975b): *The Vector Equilibrium*, lecture given as part of the lecture series *Everything I Know*. Bell Labs studios, Philadelphia. Online: *https://www.youtube.com/watch?v=9sM44p385Ws* (last access: 24 January 2016).

Gray, Jeremy (2006): *Worlds out of Nothing: a course on the history of geometry in the 19th century*. New York: Springer.

Grimm, Jacob und Wilhelm (1936): *Deutsches Wörterbuch*, vol. 11, Leipzig: Hirzel.

Hasse, Helmut (1932): *Zu Hilberts algebraisch-zahlentheoretischen Arbeiten*. In: Hilbert, David: Gesammelte Abhandlungen, Bd. 1: Zahlentheorie. Berlin: Springer, pp. 528–535.

Heath, Thomas Little (ed. and tr.) (1908a): *The thirteen books of Euclid's Elements: Vol. I, introduction and books I–II*. Cambridge: Cambridge University Press.

Heath, Thomas Little (ed. and tr.) (1908b): *The thirteen books of Euclid's Elements: Vol. III, Books X–XIII*. Cambridge: Cambridge University Press.

Heath, Thomas (1921): *A History of Greek Mathematics, vol. 1*. Oxford: Clarendon Press.

Hilbert, David (1899): *Grundlagen der Geometrie*. Leipzig: Teubner.

Kempe, Alfred Bay (1877): *How to draw a straight line: a lecture on linkages*. London: Macmillan and Co.

Kepler, Johannes (1966): *The six-cornered snowflake: a new year's gift*. Ed. and trans. by Hardie, Colin. Oxford: Clarendon Press (Latin original: Kepler, Johannes [1611]: *Strena, seu De nive sexangula*. Frankfurt-am-Main: Godfrey Tampach).

Klein, Felix (1872): *Vergleichende Betrachtungen über neuere geometrische Forschungen*. Erlangen: A. Duchert.

Klein, Felix (1893): *Vergleichende Betrachtungen über neuere geometrische Forschungen*. In: Mathematische Annalen, vol. 43, pp. 63–100 (Revised version of Klein 1872).

Klein, Felix (1897): *Famous problems of elementary geometry*. Tr. by Beman, Wooster W. / Smith, David E.. Boston et al.: Ginn and Co.

Klug, Aaron (1992): *From Macromolecules to Biological Assemblies*. In: Malmström, Bo G. (ed.): Nobel Lectures, Chemistry 1981–1990. Singapore: World Scientific Publishing Co., pp. 77–109.

Klug, Aaron (1995): *The Return of the Renaissance Man. Hugh Aldersey-Williams Talks to Aaron Klug, the Next President of the Royal Society*. In: The Guardian, 30 November, pp. 10.

Krausse, Joachim (2002a): *Buckminster Fullers Modellierung der Natur*. In: Arch+, vol. 159/160, pp. 40–49.

Krausse, Joachim (2002b): *Environment Controlling – für eine Welt der Vielen. Buckminster Fullers Wirkung in Großbritannien*. In: Anna, Susanne (ed.): Norman Foster: Architecture is about people. Köln: Museum für Angewandte Kunst, pp. 89–112.

Krausse, Joachim (2016): *Im Laboratorium von Synergetics. Buckminster Fullers Lehre vom Zusammenwirken more geometrico*. In: Petzer, Tatjana / Steiner, Stephan (eds.): Synergie. Kultur- und Wissensgeschichte einer Denkfigur. Paderborn: Wilhelm Fink, pp. 167–226.

Krausse, Joachim / Lichtenstein, Claude (eds.) (1999): *Your Private Sky: R. Buckminster Fuller. The Art of Design Science*. Zürich: Lars Müller.

Krausse, Joachim / Lichtenstein, Claude (eds.) (2001): *Your Private Sky: R. Buckminster Fuller. Discourse*. Zürich: Lars Müller.

Kroto, Harold W. (1996): *Die Entdeckung der Fullerene*. In: Krätschmer, Wolfgang / Schuster, Heike (eds.): von Fuller bis zu Fullerenen. Beispiele einer interdisziplinären Forschung. Braunschweig / Wiesbaden: Friedr. Viehweg & Sohn Verlagsgesellschaft, pp. 53–80.

Lawrence, Snežana (2011): *Developable surfaces: Their history and application*. In: Nexus Network Journal, vol. 13, no. 3, pp. 701–714.

Mancosu, Paolo (1998): *From Brouwer To Hilbert. The Debate on the Foundations of Mathematics in the 1920s*. Oxford: Oxford University Press.

Marks, Robert W. (1960): *The Dymaxion World of Buckminster Fuller*. New York: Reinhold.

Mehrtens, Herbert (2004): *Mathematical Models*. In: de Chadarevian, Soraya / Hopwood, Nick (eds.): Models: the third dimension of space. Stanford: Stanford University Press, pp. 276–306.

Morgan, Gregory J. (2003): *Historical review: viruses, crystals and geodesic domes*. In: TRENDS in Biochemical Sciences, vol. 28, no. 2, pp. 86–90.

Monge, Gaspard (1785): *Mémoire sur les développées, les rayons de courbure, et les différents genres d'inflexions des courbes à double courbure*. In: Mémoires de divers sçavans, vol. 10, pp. 511–50 (written in 1771).

Monge, Gaspard (1809): *Application de l'analyse à la géométrie, à l'usage de l'Ecole impériale polytechnique*. Paris: Bernard.

Monge, Gaspard (1811): *Géographie descriptive*. Paris: Klostermann.

Otto, Frei (1985): *Biologie und Bauen*. In: id.: Naturliche Konstruktionen: Formen Und Konstruktionen in Natur Und Technik Und Prozesse Ihrer Entstehung, 2nd ed. Stuttgart: Deutsche Verlags-Anstalt.

Pasch, Moritz (1882): *Vorlesungen über neuere Geometrie*. Leipzig: B. G. Teubner.

Pieri, Mario (1898): *I principii della geometria di posizione composti in sistema logicao deduttivo*. In: Memorie della R. Accademia delle Scienze de Torino, vol. 48, pp. 1–62.

Pressley, Andrew (2001): *Elementary Differential Geometry*. Heidelberg: Springer.

Proclus (1992): *A Commentary on the First Book of Euclid's Elements*. Ed. and trans. by Morrow, Glenn Raymond. Princeton: Princeton University Press.

Reich, Karin (2007): *Euler's Contribution to Differential Geometry and its Reception*. In: Bradley, Robert E. / Sandifer, Ed (eds.): Leonhard Euler: Life, Work and Legacy. Studies in the history and philosophy of mathematics. Amsterdam / Boston: Elsevier, pp. 479–502.

Roberts, Siobhan (2006): *King of infinite Space*. New York: Walker & Company.

Rosenfeld, Boris A. (1988): *A History of Non-Euclidean Geometry: Evolution of the Concept of a Geometric Space*. Trans. by Shenitzer, Abe. New York: Springer.

Rotman, Joseph (1999): *An Introduction to the Theory of Groups* (Graduate Texts in Mathematics, 148). New York: Springer.

Row, Sundara Tandalam (1893): *Geometrical exercises in paper folding*. Madras: Addison Co.

Row, Sundara Tandalam (1906): *Elementary solid geometry*. Trichinopoly: Saint Joseph College's Press.

Rowe, David (2013): *Mathematical models as artefacts for research: Felix Klein and the case of Kummer surfaces*. In: Mathematische Semesterberichte, vol. 60, no. 1, pp. 1–24.

Riemann, Bernhard (1868): *Ueber die Hypothesen, welche der Geometrie zu Grunde liegen*. In: Abhandlungen der Königlichen Gesellschaft der Wissenschaften zu Göttingen, vol. 13, pp. 133–150.

Sattelmacher, Anja (2013): *Geordnete Verhältnisse. Mathematische Anschauungsmodelle im frühen 20. Jahrhundert*. In: Berichte zur Wissenschaftsgeschichte, vol. 36, no. 4, pp. 294–312.

Semper, Gottfried (1983): *London Lecture of November 11th, 1853: Outline for a system of comparative Style-Theory*. Ed. by Mallgrave, Harry Francis. In: RES: Journal of Anthropology and Aesthetics, no. 6, pp. 8–22.

Semper, Gottfried (1986): *London Lecture of November 18th, 1853: "The Development of the Wall and Wall Construction in Antiquity"*. Ed. by Mallgrave, Harry Francis. In: RES: Anthropology and Aesthetics, no. 11, pp. 33–42.

Semper, Gottfried (2004): *Style in the Technical and Tectonic Arts; or, Practical Aesthetics*. Trans. by Mallgrave, Harry Francis. Santa Monica: Getty Research Institute.

Tarski, Alfred (1967): *The completeness of elementary algebra and geometry*. Paris, Centre National De La Recherche Scientifique, Institut Blaise Pascal. Reprinted in: Givant, Steven R. / McKenzie, Ralph N. (eds.): Alfred Tarski: Collected Papers, Volume 4. Basel: Birkhäuser 1986, pp. 291–346.

Tarski, Alfred / Givant, Steven (1999): *Tarski's system of geometry*. In: The Bulletin of Symbolic Logic, vol. 5, no. 2, pp. 175–214.

Ubell, Earl (1962): *Virus – a Triumph and a Photograph*. In: New York Herald Tribune, 6 February.

Wiener, Hermann (1892): *Über Grundlagen und Aufbau der Geometrie*. In: Jahresbericht der Deutschen Mathematiker-Vereinigung, vol. 1, pp. 45–48.

Wussing, Hans (1984): *The Genesis of the Abstract Group Concept: A Contribution to the History of the Origin of Abstract Group Theory*. New York: Dover.

Young, Grace Chisholm / Young, William Henry (1905): *Beginner's Book of Geometry*. New York: Chelsea publishing company.

Michael Friedman
Email: *michael.friedman@hu-berlin.de*

Image Knowledge Gestaltung. An Interdisciplinary Laboratory.
Cluster of Excellence Humboldt-Universität zu Berlin.
Sophienstrasse 22a, 10178 Berlin, Germany.

Joachim Krausse
Email: *j.krausse@design.hs-anhalt.de*

Hochschule Anhalt, Fachbereich Design,
06818 Dessau-Roßlau, Germany.

Lorenzo Guiducci, John W. C. Dunlop, Peter Fratzl

An Introduction into the Physics of Self-folding Thin Structures

Introduction

Spontaneous folding of matter has long been the subject of disparate branches of research, from the earlier vitalistic view of Leibniz[1] to the mathematical description of morphogenetic processes of D'Arcy Thompson.[2] Today, scientific advancements have shown how folding is a common strategy adopted in biological systems to build up more and more complex structures – in proteins, from a peptide chain to a functional enzyme; in plants, from bud petals to a developed flower; in organisms, from layers of cells to diversified embryos. One might think that such folding processes require exceptionally complex biological machineries to orchestrate them. On the contrary, with this contribution, we will show that folding can result from remarkably simple processes – and equally applies to both natural and artificial systems. In the following we will introduce the reader to the necessary theoretical concepts that are needed to understand these phenomena, providing examples from the common experience enabling a more systematic understanding.

When we talk about self-folding[3] thin structures we refer to a broad class of spontaneous shape changes (that is not caused by an external load) that occur in thin bodies. The adjective thin here means that a three-dimensional body has at least one dimension much smaller than the other; that is, a rod can be considered one-dimensional since it has two small dimensions; a plate instead is two-dimensional since it has one small

---

1　Deleuze 1993.

2　Thompson 1953.

3　In this document the terms layer, sheet, film are used as synonyms for plate. Similarly, folding stands for a generic out-of-plane deformation of a plate; other terms adopted for it are: bending, shape change, morphing, wrinkling, pattern formation.

dimension. In particular we will refer to bodies that can be considered two-dimensional, that is plates (which are flat) and shells (which are curved). As will become clear in section 1, the fact that plates (and shells) are thin is at the base of the rich panorama of shape changes that are observed.

The goal of the present document is to provide an overview of some of the scientific literature dealing with shape changes in self-folding thin structures. Although these shape changes can be of a very different nature, size and origin – occurring in artificial as well as biological materials, and in systems that range from nanometer to centimeter sizes – all can be understood under some unifying concepts that will be introduced in the following. Thanks to our presentation, it becomes much easier to locate where research efforts have been focused until now and to identify new fields of interest that scientific research hasn't addressed yet.

This document is organized as follows. In section 1 we introduce the reader to the basic concepts of the mechanics of a thin plate (the experienced reader can skip this part). In section 2 we will introduce many examples of self-folding systems, categorizing them on the base of how eigenstrains are distributed and orientated in the plate. In this section, we will discuss the folding (or more generally the morphing) of systems that: undergo confined expansion due to an attached elastic foundation (section 2.2); are subjected to the so-called edge growth (section 2.3); which change curvature (section 2.4). In section 3 we will give some concluding remarks on the present work.

1 Mechanics of a thin plate: basic concepts and examples

In this section we introduce the reader to the mechanics of a thin plate through some guiding examples. In section 1.1 we will analyze simple deformation modes like stretching and bending, introduce the concept of buckling of a thin body and also clarify why these two modes cannot be separated for curved shells. In section 1.2 we will present the concept of eigenstrain and eigencurvature and some of the different strategies to generate these in natural and artificial materials.

1.1 Simple deformation modes of a thin body

1.1.1 What strains are

To be able to discuss any self-folding system, we need to "zoom" in the material that make up these objects and describe what happens locally when such shape changes occur. As a guiding example, we take a sheet of rubber (such as a deflated balloon), mark it with a circle enclosing a small portion and conduct a series of three "handheld"

Lorenzo Guiducci, John W. C. Dunlop, Peter Fratzl

experiments on it (fig. 1 a−c). If we stretch it, we see that the circle becomes elongated. If we now take half of the rubber piece and stretch it by the same amount we will see that the circle is now more elongated (the elongation is now double). Stretching the sheet along a different direction will cause the circle to elongate along that direction. A useful concept to group together these intuitive results is that of strain (usually associated to the greek letter ε) in our examples the elongation of the circle with respect to its initial size. We see that the strain:

› describes the shape change of the material locally
› its magnitude can be evaluated from the ratio between the elongation and the size of the object – it is normalized
› describes the shape change along all possible directions

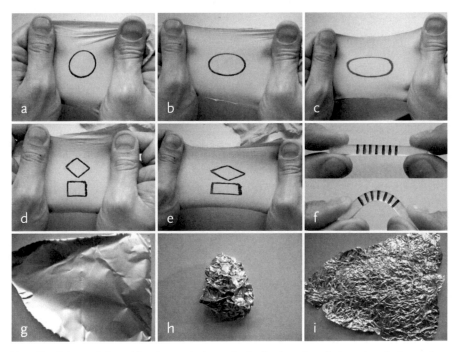

Fig. 1: Visualizing strain in a rubber sheet under increasing loading (a to c); strain depends on its frame of reference: the principal directions identify directions of pure "size-changing" strains (square to rectangle) whereas other frame orientations will give also "shape-changing" strain components (square to rhombus) (d−e); imposing a curvature to a bar generates extensional/compressive strains on its convex/concave sides (f); differently from the rubber example, an aluminum foil deformed above a small critical strain does not recover its initial shape when loads are removed (g−i).

Now let us draw two squares on the sheet, close together and in the center of the sheet, one with two sides parallel to the stretching direction and the other rotated by 45° degrees. After stretching the sheet one has become a rectangle, the other one a rhombus (fig. 1 d−e).

It seems that the first square experiences normal strains, which measure a change in length, while the other shear strains, which quantify a pure change in shape with no change in volume: this is counterintuitive because we expect that the rubber is subjected to the same strain state (at least close to the sheet center). This evidence shows that the strains depend on the frame of reference. It also shows that in each point one can find an orientation of this frame that make the shear strains null. When this happens, the strain tensor has only components on the diagonal, and they are called principal strains.

1.1.2 What stresses are

The reason why the circle elongates can be made manifest by cutting the circle out of the stretched rubber: then the circle becomes round again. If we stretch two identical overlapping sheets of rubber, we will need double the force to elongate the circle by the same amount. These evidences can be unified by a quantity – the stress, usually associated to the greek letter σ – which:

1. describes the net internal forces acting on the material locally (along all possible directions)
2. results from (and balances) the external forces of the surrounding material acting on the circle
3. is a measure of the forces acting in the material, normalized by the area of the surface on which the forces apply (in our example the area of the sheets' cross-section)

Coming back to our experiments, we notice that releasing the rubber sheet causes it to return to its initial shape, regardless of the amount of force or elongation we applied. This ability is due to the rubber's property – it is said to be *elastic*. Also, in the second experiment -stretching a half sheet by the same amount- we realize that the force we had to put on the rubber was, most probably, double. Recalling that also the strain doubled, it appears that the stress is proportional to the strain. The relationship between stress and strain is called the constitutive law of the material.[4] This constitutive law serves to define the mechanical behavior of the material, and is a function of the material's microstructure and chemical composition. In the case of rubber and many other solid materials at low strains this relationship is linear, meaning that stress and strains are related by proportionality constant. To summarize, rubber has a linear elastic behavior (at least for not too large strains): it is also said that it follows the Hooke's law.[5] Other materials (like the foil we use to wrap food, for example) would deform irreversibly when deformed above a certain strain (see fig. 1 g–i).

---

4   Callister 2007.
5   The Hooke's law describes how linear springs behave mechanically: the force needed to extend a spring increases with its extension. The proportionality factor between force and extension is the spring rigidity constant.

1.1.3 Bending of a plate[6]

Instead of stretching the sheet, we now bend it. To highlight the strain when the sheet is bent, we draw equidistant lines on the sheet's profile perpendicularly to the sheet's long axis. At the curved site, the lines are now more closely spaced on the convex side and more widely spaced on the concave side (see fig. 1f). This means that an imposed curvature[7] results in normal strains that adopt a "butterfly" distribution, negative at the convex and positive at the concave side, instead of a flat one as in the simple elongation case. Also this means that the average strains in the thickness are zero: the length of the sheet does not change, just its curvature.

When considering very thin bodies, we can make the assumption that their thickness is zero and study them as surfaces in three-dimensional space. Therefore the strain definition simplifies to only in-plane components that equal their average across the thickness. If a thin body is deformed in pure bending, under the small thickness hypothesis we would have zero energy for the bent state, because the average in-plane strain for pure bending (that is of the "butterfly" distribution) is zero. Therefore, to be able to properly describe the strain state we must define the curvature of the surface: then the imposed curvature will be associated with a strain energy. Comparing the amount of force put to bend the rubber sheet rather than stretching it, we clearly see how it is much easier to bend: the energy involved in bending is lower than in stretching, or equivalently, that the bending rigidity of a sheet is much lower than its axial rigidity. It can be shown that such mechanical behavior depends on the thin geometry of the plate: the bending energy has a prefactor h^3 and therefore will be much smaller than the stretching energy (with prefactor h) when the plate thickness h is small.

1.1.4 Buckling and stability: switching between stretching and bending modes

To understand how shape changes occur in thin bodies it is useful to introduce a simpler example of a bar subjected to an increasing axial compression. Initially the bar shortens: at every small increment of the axial load corresponds a small increment of the bar's shortening. This fact is in accordance with the symmetry of the loading case: the bar is a one-dimensional body, the load is also directed along the bar's axis. The deformation requires work to be done on the bar: this external work is stored as strain energy in the material. In other words, conservation of energy implies that we can equate this (external) energy to an (internal) strain energy that depends from the stresses and strains. Coming back to the bar loading experiment, if the load passes a certain threshold the bar will start to bend: at this point it becomes easier to bring the ends of the bar together (lower load increments are required) and every increment in load will cause a large

---

6   Most of the concepts introduced in this and the next two sections have been taken from: Audoly/
     Pomeau 2010.

7   For a definition of curvature refer to section 1.1.5.

deflection from the straight axis. This phenomenon is called buckling. It shows that for a mechanical system more than one equilibrium configuration are possible[8] (in our case, both the straight and slightly deflected ones are indeed possible at the buckling load threshold). In general the system will choose a deformation path which corresponds to lower energy: for a thin body (such as the bar but also a plate) this lower energy path is the bending mode. Moreover, since such buckling events are eventually caused by the different deformation modes that the material experiences, they can surface even in absence of external loading: such is the case of shape changes in the examples that will be provided in section 3, where buckling events are triggered by differential eigenstrains.

1.1.5 Geometry of a thin body and implications for its mechanical behavior

From a geometrical perspective, thin bodies can be considered as surfaces. To describe the shape of a surface in three dimensions, we first step back and consider, for simplicity, a line drawn on a plane. Its shape is characterized by the curvature, defined as the inverse of the radius of the circle tangent to it. For every point of the line the curvature is single-valued. If we consider a point on a generic surface instead, we can draw an infinite number of lines passing through that point. In this case each point is associated with an infinite number of curvatures depending on the orientation of the corresponding line. In reality these curvatures are related together: similarly to what we have seen for strain and stress, it suffices to know the two principal curvatures (the maximum and minimum curvatures of all possible ones), which are oriented along perpendicular directions, to fully characterize the curvature state.

Inspecting the values of the principal curvatures for different regular surfaces is instructive. Then for a plane both are clearly 0. For a cylinder[9] these are 0 and $1/R$, respectively parallel and perpendicular to a straight line drawn on its surface. For a sphere these are both $1/R$. Defining the mean curvature H as the average between the principal ones, we have a way to measure the total "amount of curvature": the sphere has a higher curvature than the cylinder, which is more curved than the plane. Another fundamental measure of curvature is the Gaussian curvature K_G, defined as the product of the principal ones (fig. 2). The meaning of K_G can be easily understood by comparing its value in the three surfaces we just presented. Plane and cylinder have same Gaussian curvature $K_G = 0$. Indeed a cylinder can be cut and unrolled to a planar state applying a pure bending deformation. On the contrary, this does not apply to a sphere. Recalling that bending deformations produces zero average in-plane strains of the surface, it now becomes clear that Gaussian curvature is related to the in-plane strains of the surface. In particular, K_G identifies a surface's metric, which is the way we calculate distances and areas on it. To grasp the concept of a surface's metric we refer to the following experiment. We draw a circle of radius r on a sphere, a plane or a saddle-shaped surface (which have respectively $K_G > 0$,

8   Gambhir 2004.
9   Pressley 2009.

$K_G=0$, $K_G<0$). We do this with the help of a thread (of length r), attaching one extremity to a point on the surface and the other one to a pen, taking care of keeping the thread in contact with the surface while tracing the circle. The length of the resulting circle will be $2\pi r$ on the plane, but lower and higher in the sphere and the saddle, respectively.

The condition of $K_G=0$ identifies a special class of surfaces called developable surfaces, that are flat up to a bending deformation. The developability condition ($K_G=0$) implies that one of the two principal curvatures is zero (recalling that K_G is defined as their product). We already have seen that a cylinder is a developable surface: in this case the direction of zero curvature is the same everywhere besides cylinders there are two more types of developable surfaces: cones and the so-called tangent developables. In each point of a cone, the zero curvature direction points to the cone's vertex. Tangent developables are the surfaces built by "extruding" a curve in the three-dimensional space along its tangent. Therefore, they are locally flat along the extrusion direction and hence developable.

$$K_G<0 \qquad K_G=0 \qquad K_G>0$$

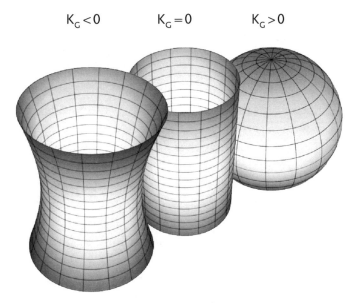

Fig. 2: A saddle, cylinder and sphere, respectively examples of surfaces with negative, zero and positive Gaussian curvatures (K_G).

From our standpoint, of studying surfaces that change shape, the Gaussian curvature has a fundamental relevance. Surfaces with same K are called isometric and (as we have seen for cylinder and plane) can be transformed to one another keeping the distances unchanged (this is the main result of Gauss' *Theorema Egregium*[10]). In general (just

10   Gauss 1902.

like what we have seen for the compressed bar) surfaces preferentially undergo some isometric shape change when subjected to a load (as we have seen thin bodies are much less rigid when bent than stretched). For developable surfaces ($K_C = 0$) such isometric transformation is easy to predict because it involves only bending. Instead, for non-developable surfaces, bending is always coupled to stretching: in this case the transformation is not isometric and therefore the shapes that can be acquired will depend strongly on the relative amount of bending and stretching energies involved, and buckling events (switching from one mode to the other) can appear when these two energy contributions have similar magnitude. Therefore, we will describe the shape change of thin bodies in terms of both the in-plane strains and the curvature-changes of their mid-surface.

1.2 Fundamentals of spontaneous shape change

1.2.1 Generating forces and deformations: eigenstrain and eigencurvature

Besides external mechanical loading, shape transformations can occur also due to other types of interaction of the body's material with the environment. For example, when we soak water beads[11] the hydrogel material of which the beads are made absorbs large amounts of water, and the beads increase in volume, undergoing large strains (identified by a zero at the subscript: ε_0, see also fig. 3a). We define this "non-mechanical" strain as the eigenstrain: it is the spontaneous or natural strain of a material which is sensitive to changes of the environment, such as temperature, light, humidity, electric or magnetic field. This character is well conveyed by the german word *eigen* which means own. The strain that we introduced previously is now called elastic strain, to highlight its mechanical origin. When both a mechanical and environmental stimuli are applied, for instance if we squeeze the swollen gel bead, the elastic strain is the difference between the total strain (this is the "visible" one that can be measured as described for the rubber sheet) and the eigenstrain; in other terms, the elastic strain is defined with respect to the swollen configuration that acts as a reference.

Even when only eigenstrains are present, residual stresses can develop and therefore elastic strains emerge. A typical example (fig. 3b) is a thin plate made of two identical layers except for different thermal expansion properties (one expands by ε_0 for a given temperature increase, the other one does not expand at all). As we could expect intuitively, when the temperature increases, the material cannot reach the local eigenstrain value because each layer constrains the other one. The expanding layer experiences a residual compressive stress while the passive layer a tensional one, being on average zero over the thickness.[12] As a result, the plate bends in order to fulfill the different expansions.

---

11   These are used to irrigate plants over longer time, for instance during short holiday trips.
12   This scenario is not always granted: some eigenstrain distributions can also prevent the plate to possess a stress-free shape (on this point, see section 2.3.1).

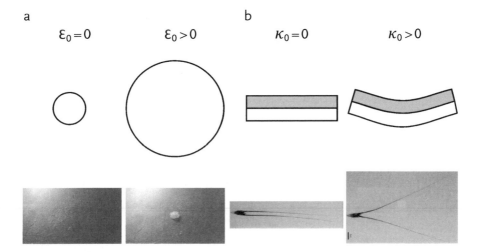

a
$\varepsilon_0 = 0$ $\varepsilon_0 > 0$ b $\kappa_0 = 0$ $\kappa_0 > 0$

Fig. 3: Eigenstrain ε_0 and eigencurvature ko are respectively spontaneous length and curvature changes. (a) Example of eigenstrain in a hydrogel: from the initial dry state ($\varepsilon_0=0$) to the final swollen state ($\varepsilon_0>0$) a large volume increase due to swelling of water takes place. (b) Example of eigencurvature in the awns of the wild wheat: each awn is a bilayer made of a hygroscopic (white) and non-expanding (grey) materials. Due to the humidity difference between night and day, the awns' shape changes from straight ($\kappa_0=0$) to bent ($\kappa_0>0$). Elbaum et al. 2007.

A natural bilayer has been described by Elbaum,[13] in the awns of the wild wheat which foster the seed dispersal. Here, the bending is caused by differential expansion between the two layers that make up each awn due to changes in humidity between night and day. Clearly, both the artificial and the natural systems can be easily described in terms of eigencurvature, that is a natural curvature that develops in the awn in absence of external mechanical loading. It becomes now clear that eigenstrain (intended as a local change of lengths in the material) is fundamental for shape changing systems: in particular a non-uniform distribution of spontaneous expansion in a body (differential eigenstrains) as in the bilayer example are among the strategies implemented in many biological self-folding systems.

1.2.2 Some strategies to generate controlled eigenstrains and eigencurvatures

Analyzing the microstructure of biological materials can provide enough insight in the design strategies that have evolved in Nature to obtain shape changing materials. A fitting example for this is wood. When trees grow they produce new wood material. An idea of this process can be obtained by looking at the many rings in the cross section of a tree's log: the tree grows in diameter by adding layers of cells to the log. Therefore this growth process is one-way: unlike animals where tissues can be reabsorbed or

---

13   Elbaum et al. 2007.

remodeled, plants lack this ability. This can be problematic because adding more material increases the weight that the tree has to sustain. Moreover additional external loading can emerge (like a condition of persistent wind). A solution is found through some specialized tissues that can expand or contract, in this way counteracting situations of unbalanced loads and securing the structural stability of trees. As explained by Fratzl et al.[14] to understand the mechanism of expansion/contraction generation it is sufficient to consider the wood cell walls architecture and composition[15] (fig. 4a–c). These cells have a prismatic shape with walls made of a swellable matrix reinforced by almost inextensible cellulose fibrils spiralling around the cells' main axis. Therefore swelling is possible only perpendicularly to the stiff non-swelling cellulose fibrils. By controlling the cellulose orientation (measured by the microfibril angle – MFA) during their development, plant cells can generate eigenstrains, and modulate them to tensile or compressive values. Such plant tissues are responsible not only for stress generation in trees but also in a variety of seed dispersal systems, such as the pine cone (fig. 4d) or the awns of the wild wheat (fig. 3b), that function in absence of any metabolic energy source. The energy needed to mechanically actuate and perform a shape change is harvested from the environmental changes in humidity – this is the reason why these systems are also called hygromorphs. Another example of biologically developed eigenstrain is found in the seed capsule of a family of plants called Aizoaceae (commonly known as iceplants). They are remarkable because -unlike most hygromorphs where eigenstrains are typically in the 10% range-, their active tissue experiences a 400% eigenstrain as a result of water absorption. The origin of such large eigenstrains is found in the peculiar composition and architecture of the active tissue, as explained by Harrington[16] and colleagues. At a microscopic scale it consists of a lignified honeycomb structure with diamond shaped cells whose walls are lined by a cellulose rich hygroscopic tissue (fig. 4e). As liquid water comes in contact with the tissue, the hygroscopic inner layer swells increasing its volume; the honeycomb structure directs this large volume increase in a large expansion along the cells' short axis. Man-made materials are also subjected to eigenstrains. Hydrogels in particular are a special class of polymers undergoing large expansion (1000% in volume) upon water sorption. Hydrogels have a chemical structure that can be described as a network of long chains connected together at points called cross-links. These chains are hydrophilic (water liking) therefore they "attract" water molecules when soaked. The degree of expansion depends on the density of cross-links: less cross-links bring to higher expansion. Moreover, they can be chemically synthesized to modulate the swelling in response to a variety of different stimuli (temperature, light, magnetic and electrical fields). Technological advances have made possible to fine tune both the stimuli-sensitivity (for instance one

---

14   Fratzl/Elbaum/Burgert 2008.
15   Composition refers to the chemically different materials found in the tissue; architecture refers to the structure of the tissue at lower scales (microstructure, nanostructure, etc).
16   Harrington et al. 2011.

can control the temperature at which volume change occurs) and the spatial distribution of the cross-links in thin geometries (by lithographic techniques as shown, for example, by Klein and Kim,[17] see fig. 4f). For these reasons, hydrogels have been the experimental system of choice to investigate the role of eigenstrains distribution onto the folding shape.

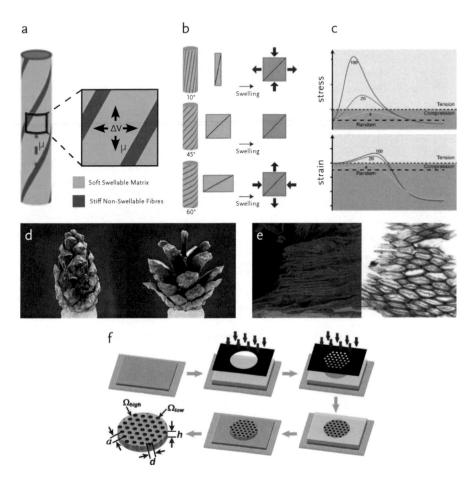

Fig. 4: Strategies to control eigenstrains in biological and artificial systems. (a–c) The cellulose microfibril angle (MFA) controls the mechanics and swelling of the plant cell wall. Dunlop 2016. (a) Schematic of a simplified cell wall, consisting of a soft swellable matrix reinforced by stiff fibres wrapped at an angle μ around the cell. (b) Schematic of how cells with different MFAs will respond to swelling of the soft matrix. (c) Tensional and compressive strains and stresses can be generated depending on the MFA. (d) A pine cone, an example of actuating biological structures responding to change in environmental humidity based on the plant cell wall model. (e) Larger directed expansions are achieved in the active tissue of the iceplant seed capsule through a different spatial arrangement of the swelling (pale blue) and non-swelling (red) materials. (f) Schematic of halftone lithography: by changing the dimension of the low swelling black dots with respect to the high swelling green material fine spatial control of swelling properties in a hydrogel sheet is possible. Kim et al. 2012.

17   Klein/Efrati/Sharon 2007; Kim et al. 2012.

1.2.3 Minimal triggers for spontaneously morphing systems

From what we have seen in this section, in order to achieve a shape change a self-folding system must possess two basic ingredients:

> ability to deform in response to a non-mechanical stimulus (eigenstrain)
> non-uniform distribution of eigenstrain across the body

Especially this second point is very relevant, as we will see, because the eigenstrain distribution influences which pattern of shape change arises. In reality the choice of this pattern or the evolution from one pattern to another is strictly subjected to the so-called boundary conditions. Mathematically speaking these are additional equations that have to be considered when searching for the equilibrium shape of a self-folding body. From a more practical perspective, these are physical constraints acting on the body: for instance, in the example of the spontaneously bending bilayer, boundary conditions describe how the extremities of the bilayer are treated (for example, if an extremity is glued to a support, and the other one is free to move along a carriage).

2. Exploring the manifold of thin structures undergoing spontaneous folding

2.1 A classification of self-folding systems based on eigenstrains distribution

To properly distinguish among the various types of biological and artificial self-folding systems, we choose to categorize them on the basis of the eigenstrain distribution and orientation. Moving along the vertical axis of fig. 5, a first distinction is made between systems that behave as monolayer or bilayer systems, that is systems that are respectively homogeneous or not across the thickness (first or second row of examples in the fig.). From what we have shown in section 1.2.1, this is equivalent to endow the surface with an eigenstrain or eigencurvature. A second distinction is made depending on the in-plane direction and distribution of the eigenstrain (moving along the horizontal axis): the eigenstrain can expand/contract the layer mostly along one direction (and be null on the perpendicular one) or all directions (isotropic); they can be uniform in the whole layer or be differently distributed in the layer (for instance concentrated at the edge). In the following, eigenstrain and eigencurvature will be expressed in their principal frame of reference as vectors $\varepsilon_0 = [\Delta_1 \ \Delta_2]$ and $\kappa_0 = [\kappa_1 \ \kappa_2]$. In this notation, the subscripts 1, 2 refer to the principal directions of the sheet (both flat and curved), indicated by the axes 1, 2 in fig. 5.

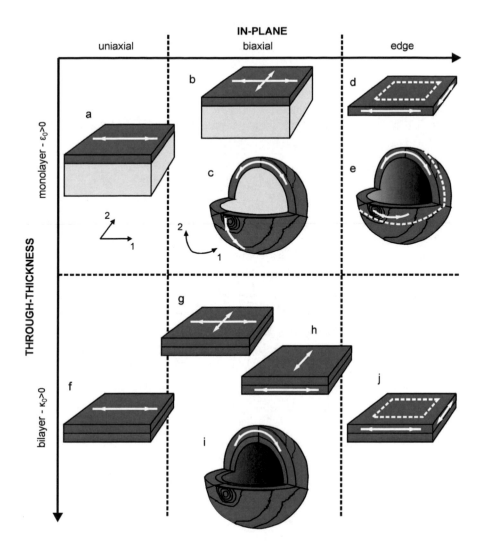

Fig. 5: Different types of self-folding systems triggered by in-plane (horizontal axis) and through-thickness (vertical axis) distribution of eigenstrains. Top row, from left to right: monolayer systems characterized by a uniaxial, biaxial and edge confined eigenstrains distribution. Bottom row, left to right: bilayer systems characterized by single, double and edge confined eigencurvature (eigencurvature derives from a through-thickness distribution of eigenstrains). The principal values of eigenstrains and eigencurvatures are directed along the white arrows. The geometry of the system (flat or doubly-curved) as well as the boundary conditions to which it is subjected (pale blue material schematizes a soft elastic foundation) have strong influence on the folding process.

2.2 Soft matter with hard skin

In this section we present some examples of shape-changing systems formed by a thin and stiff sheet attached to a soft substrate (respectively schematized in fig. 5a–c by dark blue and pale blue materials). In all these cases the shape change is due to a differential strain between these two portions: this gives rise to a manifold of different shapes and patterns.

2.2.1 Shape changes in a sheet subjected to uniaxial confinement

To start off we consider the shape transformation of a thin sheet that is attached to a soft substrate. This is a common schematization of many systems that appear at disparate length scales, both in biological and artificial systems,[18] such as the shape of a polyester sheet resting on a liquid or a soft hydrogel (mm scale), or tri-colloidal systems and even surfactant-water-air interface in the lungs alveola (nm scale). All these systems start from a flat state that subsequently morphs into a wavy (wrinkled) configuration and eventually reach a convoluted, folded one. The morphing can be caused by a lateral compression of the sheet or, equivalently by a unidirectional expansion of the sheet subjected to a lateral confinement: both cases can be described by defining the normalized excess length of the sheet (fig. 5a). In mathematical formalism the sheet is subjected to an eigenstrain $\varepsilon_0 = [\Delta \ 0]$ with $\Delta > 0$ since the sheet expands with respect to the substrate.[19] The progression of the shape change depends on the magnitude of Δ as we will show hereafter.

Ideally, compressing the sheet by Δ should shorten while keeping it flat. Instead, even for very minor compression, it starts wrinkling (fig. 6a). The wrinkling is the onset of a first mechanical instability which is in fact very similar to the buckling of an axially loaded slender bar: from our experience we know that it is very difficult to shorten the bar, which instead will preferentially deform by bending. The shape of the wrinkled sheet is a sine wave with a profile length λ_0, where λ_0 depends both on the bending rigidity of the sheet B and the effective stiffness of the substrate K. This can be readily understand this if we first consider the sheet separately from the substrate. If we would compress the sheet alone, it would bend acquiring a single arched shape, pretty much like as for the axially loaded bar: this is because a single large arch has a smaller curvature than many smaller ones. On the other hand, a single arch would cause a large deformation of the substrate, since the latter is bonded to the sheet and follows its vertical displacement. Therefore the substrate switches to a configuration with many arches. Eventually the actual length of the sine profile λ_0 is a tradeoff between these two competing processes.

---

18 Pocivavsek et al. 2008; Genzer/Groenewold 2006; Brau et al. 2013.
19 With reference to fig. 5a, considering a principal reference system aligned with the sheet sides, Δ is the eigenstrain component along the horizontal direction (marked by a white arrow), whereas the component along the perpendicular direction is zero. This notation is explained in section 2.1.

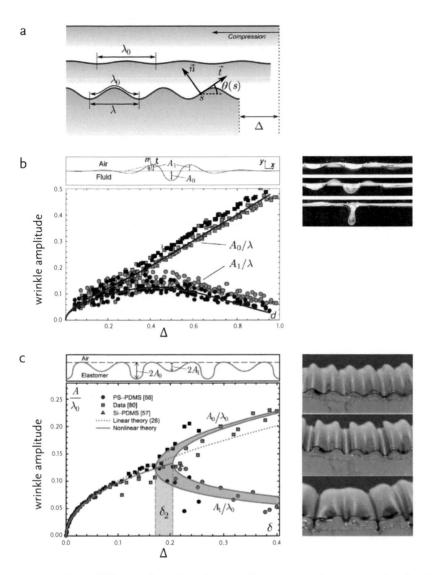

Fig. 6: Wrinkling and folding in a sheet subjected to uniaxial eigenstrain Δ with respect to an elastic foundation. (a) The initial straight state is lost in favor of a periodic wrinkled pattern. (b) Wrinkle-to-fold transition when the foundation is a liquid: past a certain threshold, the central fold amplitude A_0 grows at the expense of flanking wrinkles A_1. (c) Wrinkle-to-fold transition when the foundation is a soft solid: flanking wrinkles do not disappear and the monolayer acquires a typical periodic profile with alternating deep folds and shallow wrinkles. Brau et al. 2013; Pocivavsek et al. 2008.

As Δ increases the wrinkled configuration is not stable anymore: this can be observed in the plot of fig. 6b, showing the measured amplitudes of the central wrinkle A_0 and the first flanking one A_1 taken from a plastic foil floating on a liquid. When Δ increases until about one third of the profile length λ_0, the wrinkled profile starts changing and the center wrinkle amplitude A_0 starts increasing linearly at the expense of the flanking ones. This bifurcation point is the onset of the wrinkle-to-fold transition. In the limit of larger and larger confinement, the center fold increases and all other wrinkles disappear. In reality this is not observed because the fold contacts itself when $\Delta = \lambda_0$

The reason why the fold appears depends on how the strain energy of the system scales with an increasing confinement Δ:[20]

$$U \sim (BK)^{1/2} \Delta - K\Delta^3$$

the first term, with degree one half, is related to the winkling, and, as said, depends on both the sheet bending rigidity B and substrate stiffness K; the second term with degree three is related to the folded state and depends only on K. When Δ is small the cubic term is negligible and the winkling solution is the most stable state. When Δ increases the cubic term lowers the total energy, favoring the folded state.

A slightly different folded configuration occurs when the substrate is a soft elastic material such as an elastomer (fig. 6c). Again in this case, when Δ is small the wrinkled shape is stable.[21] Then, at a certain confinement δ_2 the wrinkle-to-fold bifurcation is observed. In this case though, there are many deeper folds of amplitude A_0 flanked by smaller wrinkles of amplitude A_1 which don't vanish even for very large Δ. This behavior is due to the elastic response of the substrate: whereas in the sheet-fluid system the sheet could freely flow on the fluid and the central fold grow at the expense of the other wrinkles, here such a horizontal deformation of the substrate has a "cost" in terms of strain energy. As a result the folded configuration has a typical period-doubling shape (see fig. 6c) with alternating small amplitude wrinkles and large amplitude folds. If even larger confinement Δ is applied a period-quadrupling folding occurs: this suggests that a cascade of spatial period-doubling bifurcations could happen at high enough confinements (more than period-quadruple foldings however are hard to be observed experimentally because the sheet can undergo self-contact).

2.2.2 Shape changes in a sheet subjected to biaxial confinement

As an extension of the previous uniaxial strain states we now consider a biaxial compression or, equivalently a biaxial growth of the sheet. This scenario is a good approximation of what happens when our skin ages. Aging triggers biological modifications that make the skin superficial layer (epidermis) stiffer, while the internal layer tends to lose water causing its shrinking. This results in an excess length of the epidermis: mathematically the sheet is subjected to an in-plane eigenstrain $\varepsilon_0 = [\Delta_1 \ \Delta_2]$ with respect to the substrate (depicted by two perpendicular arrows in fig. 5b). We know from our daily lives that -from the unaesthetical orange peel appearance, to deep wrinkles- skin blemishes can be

---

20   Pocivavsek et al. 2008, 915.

21   The amplitude is proportional to the square root of Δ as with a fluid substrate.

manifold. Moreover this problem has more than a merely aesthetical interest. Indeed a similar surface patterns appear in artificial material systems, for example depositing a thin metallic film[22] onto an elastomer[23] and letting the system cool down at room temperature, or by surface-modifying of polymers such as poly-dimetilsiloxane (PDMS) (see the examples described later).

In this case the buckling analysis has to account for all possible pairs of values Δ_1, Δ_2. Audoly and Boudaoud[24] started studying this problem for values close to the critical[25] threshold value Δ^C. They showed that the sinusoidal wrinkled configuration (also called cylindrical) of frequency k arising as the first buckled configuration in the monodimensional case is still a solution when the interfacial strain is small (below the critical value). For slightly larger eigenstrains Δ three additional possible patterns arise: the so-called, undulating, varicose and checkerboard patterns (fig. 7a). The undulating is obtained from straight wrinkles by lateral undulations of the crests and valleys. The varicose corresponds to a modulation of the amplitude of the straight wrinkles, along its crests and valleys. The checkerboard is obtained by superposition of two perpendicular sets of straight wrinkles. The hexagonal is obtained by superposition of three sets of straight wrinkles, at $2\pi/3$ angles.

These patterns are perturbed configurations of the cylindrical one, with respect to which a periodical perturbation of frequency k' along the longitudinal direction is superposed. Putting $q = {}^{k'}/_k$ as the longitudinal wavevector relative to the cylindrical one, Audoly and Boudaoud present the stability of these patterns as a function of the two differential strains Δ_1, Δ_2 and their wavelength. In the phase diagram of fig. 7b each portion of the plane is marked with the pattern that is more stable for the corresponding range of values of the two differential strains Δ_1, Δ_2. In particular they showed that the undulating pattern with large wavelength ($q=0$) are more stable than the wrinkled state as soon as the strain compressing the wrinkling pattern longitudinally is higher than the critical value $\Delta_2 > \Delta^C$. The transition from wrinkled to varicose pattern is not observed because it takes place at strains that are much larger than for undulating stripes or for the checkerboard mode. When the differential strain is equibiaxial ($\Delta_1 = \Delta_2$), the checkerboard pattern is the more stable state above the critical thresholds, especially for larger Δ. Remarkably, the equibiaxial case close to the initial buckling threshold ($\Delta_1 = \Delta_2 = \Delta^C$) is a special situation where the straight wrinkled state can coexist with all the other patterns. This is because all of the patterns (theoretically extrapolated or experimentally reported) can be seen as a superposition of straight wrinkles at different angles.

---

22 Here and throughout the rest of the text the term film is synonym of thin sheet.
23 Bowden et al. 1998.
24 Audoly/Boudaoud 2008a.
25 The critical threshold is the value of Δ_1, Δ_2 beyond which the system switches to a buckled state.

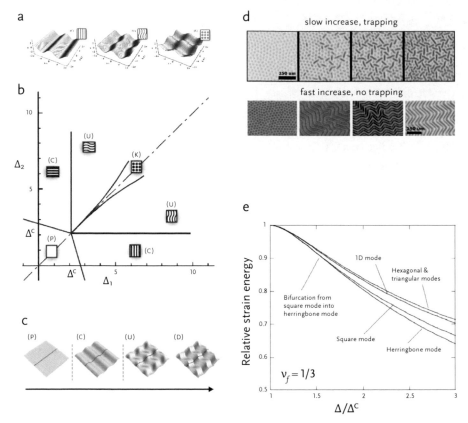

Fig. 7: Complex pattern formation and evolution in a monolayer subjected to increasing biaxial confinement resting on an elastic foundation. (a) Cylindrical, undulating and checkerboard patterns. (b) Phase diagram of the most stable patterns depending on the extent of eigenstrain along two perpendicular directions. Audoly/Boudaoud 2008a. (c) Evolution of pattern in case of an increasing equibiaxial confinement: planar, cylindrical, undulating and developable patterns. Audoly/Boudaoud 2008b. (d) Influence of rate of change of confinement onto the pattern evolution: slow increase cause "trapping" from hexagonal pattern into disordered labyrinthine one; fast increase enable reaching the more stable herringbone pattern. (e) Relative strain energy of different patterns at increasing biaxial confinements: lower energies mean more stable patterns. Cai et al. 2011.

When the differential strain increases (see fig. 7c), the undulating pattern evolves smoothly (without a well-defined buckling threshold) towards a developable surface obtained by folding a cylindrical shape along sinusoidal ridges (pattern D in fig. 7c). This pattern (derived theoretically) is very close to the so-called herringbone (or zig-zag) pattern observed experimentally. In this pattern the spacing between the ridges along the transversal and longitudinal direction (with respect to the initial straight wrinkling) are comparable. By studying the energy of the developable pattern at very large differential strains, Audoly and Boudaoud showed that this configuration is far from the global minimum of energy.[26]

---

26 The global minimum is obtained when the spacing between ridges is much smaller than the wavelength of these sinusoidal ridges.

This means that experimentally the system is trapped in a non-optimal state (herringbone pattern with comparable transversal and longitudinal wrinkling periodicity) although a more stable configuration could be possible. This so-called trapping mechanism depends on the history of loading of the system: if the differential strain beyond the secondary buckling threshold (from straight to undulated wrinkles) is increased slowly, the developable pattern inherits the wavelengths typical of the undulating pattern because this is a smooth change; if, instead, the change is fast, the system can reach the global minimum. The idea that loading history of the system has significance in the shape selection is also crucial in self-folding bilayered systems. Finally, at even larger Δ the developable pattern smoothly tends to a Miura-ori folding pattern.[27] Again since this is a smooth transition trapping is possible.

Trapping in real experimental systems (as can be seen in fig. 7d) has an utmost importance because it constitutes a way to select different pattern under similar experimental conditions. For example, Cai and Hutchinson[28] were able to show this in specimens of poly-dimethylsiloxane (PDMS) substrates that were treated with a chemical oxidative process that cause the stiffening of a thin superficial layer. These PDMS samples were then exposed to ethanol vapor: the ethanol would be absorbed first in the thin layer thus generating an equi-biaxial differential swelling. Moreover, by controlling the duration of exposure and the concentration of ethanol in the vapor, they were also able to control the speed and amount of differential swelling. Being able to inspect how the system would behave under a controlled equibiaxial differential strain $\varepsilon_0 = [\Delta \ \Delta]$, they compared the pattern formation due to a large increase in Δ, with slow and fast dynamics. With slow dynamics the system started from a pure hexagonal pattern, then isolated hexagons would coalesce with neighbors producing an extended local groove. Such groove formation was also triggering the coalescing of a neighboring pair into another groove that was not parallel to the original pair, in order to accommodate the local stress in an equi-biaxial manner. The overall result of such a sequence of individual buckling events was a pattern that locally resembles a segmented labyrinth, or a herringbone pattern with no global orientation (as observed by Huang and Lin[29]). In contrast, with a fast increase of domains of herringbone pattern were developed from the beginning and subsequently refined in extremely well-defined and larger domains.

To understand such a behavior, they derived upper bounds for the theoretical strain energy at low Δ of different patterns: the straight wrinkled, the checkerboard, hexagonal and herringbone. While close to the critical threshold all different patterns have similar

---

27 Audoly/Boudaoud 2008c; a Miura-Ori folding pattern is a traditional folding in artistic origami, where the creases are zig-zag lines oriented as the crests and valleys of the developable pattern of fig. 7c.

28 Cai et al. 2011.

29 Huang/Hong/Suo 2005; Lin/Yang 2007.

energies (see fig. 7e), these become more separated at higher levels of differential eigen-strains. Their predictions clearly show that for higher Δ the more stable (lower energy) patterns were the checkerboard (square) and herringbone one followed by the straight wrinkled and hexagonal ones. While this result confirms the same conclusions drawn by Audoly and Boudaoud, it was striking that the theory could not explain the system's dominant preference for the hexagonal pattern when D was increased slowly. To explain this oddity the basic assumptions of the theory were reconsidered. In particular the authors questioned the assumption of a perfectly flat film. Indeed before any pattern formation, the films would show (even if very minor) an initial curvature. In fact, by introducing in the theory an outward curvature of the film, it could be shown that the hexagonal buckled mode is the more stable configuration (lower elastic energy minimum) in presence of a positive spherical curvature of the film. Moreover the curvature introduces a clear asymmetry in the buckling response: this was confirmed by the evidence that the hexagonal patterns on PDMS always showed inwardly oriented dimples.[30]

2.2.3 Shape changes in a curved sheet subjected to equibiaxial confinement

These results show how important the surface curvature is in the selection of a specific pattern. Moving on from this point, many researchers focused on the effects of curvature both theoretically[31] and experimentally[32] when the curved sheet is subjected to an homogeneous in-plane eigenstrain $\varepsilon_0 = [\Delta \ \Delta]$ (see fig. 5c). The main outcome of these studies is that the surface curvature acts as an additional parameter to control the pattern formation: large changes of the film curvature (at a given level of differential strain) can push the system to a different pattern, exactly as does the critical strain far above the critical value. From the phase diagram in fig. 8a, it can be seen that the hexagonal pattern is stable when the differential strain Δ and/or the radius of curvature R are low. Increasing R (relatively to the shell thickness h) or Δ produces a so-called labyrinthine pattern with domains of curved ridges and valleys. Intermediate values of these parameters produce configurations of stable coexistence of the aforementioned patterns. At a closer inspection, these patterns are not trivial. Intuitively we would expect that for a very large radius of curvature.[33] The film assumes the herringbone pattern observed by Cao and Hutchinson instead of the labyrinthine one (see fig. 8a). But the planar periodicity of the herringbone pattern is incompatible with a spherical domain. Similarly, the hexagonal pattern observed at low radius is not perfect: some dimples have only 5 neighbouring dimples, demonstrating the existence of pentagonal topological defects (disclinations) as required by Euler's polyhedral theorem.[34]

---

30   A symmetric buckling would equally produce inward or outward dimples; instead the experiments showed only inwardly oriented dimples: in fig. 7d the inward dimples have a darker shade of grey.

31   Cao et al. 2008; Li et al. 2011; Stoop et al. 2015.

32   Breid/Crosby 2013.

33   That is in the limit of a flat film.

34   From Euler polyhedral theorem in the sphere there must be 12 dimples with 5 neighbors.

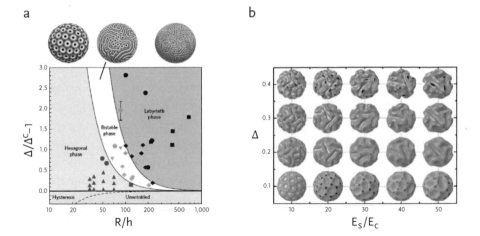

Fig. 8: Influence of curvature of the monolayer on the selection of buckling pattern. (a) Theoretical phase diagram and relative patterns (hexagonal, bistable and labyrinthine) as a function of surface curvature (normalized by monolayer thickness: R/h) and in-plane eigenstrains (beyond the onset of buckling: $\Delta/\Delta^C - 1$. Points correspond to real experiments on artificial system. Stoop et al. 2015. (b) Pattern formation on the skin of drying peas and relative numerical predictions as a function of skin-core modulus ratio and confinement. Li et al. 2011.

The wrinkling at the surface of a drying pea is a perfect system to observe such interplay between pattern formation and surface curvature (fig. 8b). They have an almost spherical shape and, when drying, the softer inner core shrinks more rapidly than the stiff outer shell: also in this case there is a positive differential strain Δ between shell and core. Typically, due to the high shell thickness and the high stiffness ratio between shell and core, the dimples that are formed are quite large compared to the pea size. Therefore when Δ increases past a critical threshold a typical buckyball pattern appears, which shows a clear arrangement of hexagonal-pentagonal dimples. At even larger Δ neighboring dimples coalesce together forming a groove, in a secondary wrinkle-to-fold bifurcation. In this case since the groove length is comparable with the sphere radius, a true labyrinthine pattern is unfavored and grooves show a scattered orientation.

2.3 Torn plastic bags and growing leaves: shape changes due to edge growth

In this section we present some examples of systems where the shape change is caused by a positive differential strain of the edge with respect to the core of the sheet (see fig. 5d–e). As the reader could already understand from the title of this section, this scenario has a very broad relevance, since it is applicable to a variety of different systems, both biological and artificial.

2.3.1 Non-uniform in-plane eigenstrain distributions

All the examples introduced until now show that surface curvature, eigenstrains magnitude and to a certain extent also speed of deformation are possible means to control the pattern of shape change. As we have seen, such patterns consist of out of plane deflections from the sheet initial configuration that can be considered large in comparison with the sheet thickness, but are small with respect to the sheet size. Another line of research investigates large shape changes in finite size sheets through a reasoned distribution of the in-plane eigenstrain. This forward problem (given a three-dimensional target shape, find the corresponding in-plane eigenstrain distribution) has been dealt with for ideal surfaces (with vanishing thickness): the field of map projections provides a number of well-established conformal mappings [35] of three-dimensional shapes onto flat surfaces (for example it is possible to almost completely map a sphere to a flat surface with the exception of a point).

The reverse problem, from a geometrical perspective, is to find the three-dimensional shape of a thin body corresponding to a given distribution of in-plane eigenstrains. This problem (as realized by Efrati, Sharon and Kupferman [36]) is not trivial and has been solved analytically only under simplifying axisymmetry assumptions. [37] A thin body loses its intrinsic geometry as a result of a generic growth process. For example, let us assume, for simplicity, the thin body is flat: the growth process leads to an in-plane eigenstrain distribution. If the elastic plate is capable of satisfying the prescribed metric, then there exists a stress-free configuration, which is unique. That is the prescribed metric transforms the flat body to a curved one. If this is not possible, a residual in-plane stress will arise in the deformed plate, regardless of the assumed shape. Efrati and coauthors called such bodies non-Euclidean to highlight their impossibility to obtain a stress-free shape. Here non-Euclidean means that their internal geometry (prescribed by the metric) cannot be described only in terms of distances between points since such a pure geometrical description would overlook the presence of residual stress. In mathematical terms it is said that such bodies are not immersible in the three-dimensional Euclidean space, or, alternatively, the prescribed metric has no embedding in the Euclidean space.

2.3.2 Edge growth in a flat sheet

To explain such reverse problem, we once again resolve to a hand-held experiment, where we take a piece of plastic bag (flat in its resting configuration) and tear it apart (fig. 9a). A travelling crack is formed: the material flows perpendicularly to the crack as a result of the large extensional stresses. These material deformations close to the newly formed edges are not recovered after the load has been removed: the material has deformed

---

35   Ivanov/Trubetskov 1995.

36   Efrati/Sharon/Kupferman 2009a; Efrati/Sharon/Kupferman 2009b; Efrati/Sharon/Kupferman 2011; Efrati/Sharon/Kupferman 2013.

37   Dias/Hanna/Santangelo 2011.

plastically. Hence we have prescribed a unidirectional eigenstrain along the edges. To be concise let us name this type of eigenstrain distribution as edge growth. The evident consequence of this can readily be seen: the edges have now assumed a rippled shape. Strikingly similar shapes are also observed in the shape of salad and kale leaves (fig. 9b): also in this case the edges grow more than the center of the leaf. The question is then: can we predict the edge shape corresponding to a given edge growth profile? Sharon[38] et al. analyzed the shape of such torn plastic sheets and realized that the wrinkles at the sheet's edge are self-similar (or fractal): each bigger wave contains along its profile 3.2 smaller waves sharing the same amplitude-to-wavelength ratio, with wavenumbers according to a sequence:

$$k, \alpha k, \alpha^2 k, \alpha^3 k, \ldots$$

where $\alpha = 3.2$ and k being the wavenumber of the first largest wave. Measuring the contour length of the sheet close to the edge, they could observe a contour excess length (with respect to the flat contour length) with a rapid decay moving from edge to center of the sheet (fig. 9c) which is mathematically equivalent to an eigenstrain $\varepsilon_0 = [\Delta \ 0]$ aligned with the sheet's sides (see fig. 5d). That is, each narrow strip parallel to the edge is in-plane compressed by the flanking one closer to the center. Recalling section 2.2.1, this is compatible with a straight wrinkling pattern with wavelength λ (depending on the sheet stiffness and thickness). Therefore the cascade of waves is due to the superposition of many such wrinkling patterns with increasing wavenumber $\alpha^i k$ as one moves closer to the edge. Then the question arises: why is the cascade scaling factor about 3.2?

Audoly and Boudaoud[39] showed that such period tripling waves are the most stable (lowest energy) types of wrinkling that can be achieved in a plate with a growing edge. To achieve this conclusion, he first realized that growth profile on the edge prescribes a given Gaussian curvature to the plate, on which the stretching energy depends. To minimize the elastic energy, the sheet assumes configurations that respect this prescribed curvature. In the limit of an infinitely thin plate, two families of embeddings (that is configurations respecting the growth profile) are possible, either with a unique wrinkling frequency or a cascade of superposed wrinkles. Because of the mathematical form of the elastic energy, its solutions must be invariant to a sign change.[40] This means that the "self- similarity factor α must map odd integers to odd integers, [leaving] only odd integers (3, 5, etc.) as eligible values".[41] Among these, fractal wrinkling patterns with $\alpha = 3$ yields the most stable solutions and therefore are chosen.

---

38   Sharon et al. 2002.

39   Audoly/Boudaoud 2003.

40   The sign invariance mentioned here expresses the equivalence of two wrinkles which are specular with respect to the sheet plane.

41   Ibid, 3.

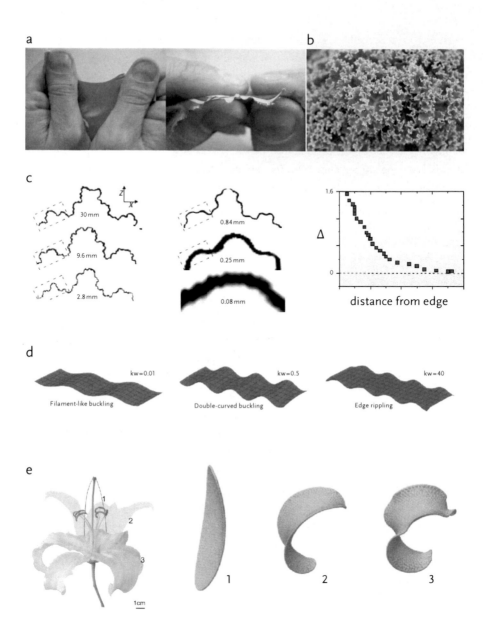

a

b

c

Z
X

30 mm

9.6 mm

2.8 mm

0.84 mm

0.25 mm

0.08 mm

1.6

Δ

0

distance from edge

d

kw=0.01

Filament-like buckling

kw=0.5

Double-curved buckling

kw=40

Edge rippling

e

1
2
3

1 cm

1

2

3

Fig. 9: Edge rippling induced by edge growth in monolayer systems. (a) When we tear a plastic bag, ripples arise at the newly formed edge which are strikingly similar to those observed in kale and salad leaves (b). (c) Self-similar edge wrinkles in a torn plastic bag and measured growth strain. Sharon et al. 2002. (d) Influence of finite sheet size on the type of edge ripples in monolayers with "blade" geometry. Liang/Mahadevan 2009. (e) The blooming of a lily results from edge growth and its doubly curved geometry: in a first stage edge growth causes curvature inversion of the petal, that is blooming (1 to 2) and at a later stage ripple formation (2 to 3). Liang/Mahadevan 2011.

2.3.3 Edge growth in narrow flat sheets

Liang and Mahadevan[42] studied a special case of edge growth in a narrow plate, a scenario that is more similar to what happens in algal blades and lily leaves.[43] Whereas in the previous examples of a torn plastic bag or a salad leaf the depth of the growth profile was much smaller than the sheet size, in these cases the finite width (w) of the plate has an influence on the type of shapes that can be achieved: algal blades can have both a saddle shape characterized by a negative Gaussian curvature everywhere or a flat configuration with rippled edges. To unravel the process of shape selection in growing blades, the authors calculated how bending and stretching energies scale in three possible configurations in presence of a large edge growth Δ confined close to the edge. The onset of a buckling mode happens when a critical growth Δ^C is reached, where bending and stretching energies are in the same order of magnitude. Their analysis showed that three types of periodic buckling modes can arise depending on the ratio between the ripple wavelength $1/k$ (itself depending on the max value of Δ) and the blade width w. In particular (fig. 9d), for $kw \ll 1$ a one-dimensional buckling would occur where the blade wrinkles along the longitudinal direction and stays flat along the transversal one; for $kw \sim 1$, the blade undergoes a doubly-curved buckling where the longitudinal wrinkling is accompanied by transversal curvature of opposite sign (a sequence of saddles); and finally an almost flat blade with an edge rippling of large wavenumber k is favored when $kw \gg 1$.

2.3.4 Edge growth in curved sheets

The blooming process of a flower has similarities with the morphogenesis of leaves but it is worth considering since it sheds light on the role of initial shell curvature in the shape change process (fig. 5e). The traditional explanation is that the central midrib is responsible for the blooming of the lily: the leafy half of the midrib, growing more than the woody half, would generate the required outward bending. By excising the midrib from the petals when the lily is in the bud stage, Liang and Mahadevan[44] observed that still an almost physiological blooming could occur. Moreover they observed ripples at the petal edges typical of edge growth, both in the bud and bloomed state. These evidences lead them to postulate an blooming mechanism driven by edge growth. In accordance to their measurements, they considered an edge growth Δ reaching farther towards the center of the petal (differently from the previous case of the "blade" leaves). Moreover, natural curved petals have a positive Gaussian curvature with a weak longitudinal curvature and strong lateral one, their ratio being $m = \kappa_{x0}/\kappa_{y0} \in [0,1]$. This is a major departure from previous examples of edge growth in flat systems: as we have seen in section 1.1.5, in a surface with non-zero Gaussian curvature bending and stretching are coupled. As the growth strain increases from zero (fig. 9e), the shell unfolds slowly decreasing both

---

42   Liang/Mahadevan 2009.
43   Since only the shape of the plate changes, we still refer to an eigenstrain distribution as in fig. 5d.
44   Liang/Mahadevan 2011.

curvatures. Past a certain critical value, the lower longitudinal curvature reverses, that is, the petal blooms. The theory shows that for the typical petal shape where κ_{x0} is low (that is $m \ll 1$), the curvature reversal is not sudden and the petal opens smoothly. Although no real buckling occurs (the blooming happens smoothly), the critical value has still a mechanical significance: it separates the stretching dominated and bending dominated regimes. Indeed after that value is passed, pure bending deformations in the form of edge-localized ripples arise.

2.4 Systems with spontaneous curvature

In this section we will give an overview of thin bodies that change shape as a result of a differential expansion across the thickness: this situation is equivalent to consider them as simple surfaces with a given natural (or eigen-) curvature and is schematized in the cases of fig. 5f–j.

2.4.1 Developable sheets with a single eigencurvature
In the most simple case, we consider a sheet which has at least one vanishing principal curvature, that is the sheet has a developable shape (as in fig. 5f where the sheet is flat). If an eigencurvature is prescribed, the sheet simply fulfills it through a bending deformation. This is the case, for instance, of the wild wheat awns (fig. 3b) and the pine cone in (fig. 4d) where the eigencurvature is caused by a differential hygroscopic expansion through the thickness. We will not pause longer on this class of morphing since no other shape changes occur when the eigencurvature increases. Indeed, since the sheet is developable, bending and stretching modes are decoupled: therefore a larger eigencurvature will only result in larger bending of the sheet.

2.4.2 Sheets with a non-developable initial or target shape[45]
The situation gets more complicated if either the initial shape or target shape are not developable (see fig. 5g–i). A typical example of the first type (non-developable initial shape, fig. 5i) is the manifold of shapes that cells can acquire (from spherical to elongated to biconcave shape typical of red blood cells as in fig. 9a). Cells' membranes are vesicles made of a bilayer of free flowing lipid molecules. This makes them quite similar to soap bubbles. On the other hand, the two monolayers do not exchange lipid molecules and therefore area differences between them can give rise to spontaneous curvature. These two aspects influence the kind of shape to be acquired. Firstly the surface tension which (similarly to soap bubbles) tends to minimize the membrane area for a given volume: this

---

45   The target shape is the shape that is obtained (if possible) by fulfilling both the eigenstrain and eigencurvature. For example, a non-developable target shape is the saddle, corresponding to prescription of equal and opposite principal eigencurvatures to an initially flat sheet.

drives towards spheres. The second is the spontaneous curvature: the larger the spontaneous curvature is, the more the vesicle will tend to satisfy it. Therefore surface tension and spontaneous curvature lead in general to contrasting shapes. Typically, spheres tend to buckle into prolate (elongated) and oblate (discoid) ellipsoids: from these shapes Helfrich,[46] Seifert and Lipowsky[47] together with Svetina and Zeks[48] studied the types of possible evolutions as an increasing spontaneous curvature is ascribed. The shapes that were found could be placed in a geometric phase diagram with axes v relative volume and c_0 relative spontaneous curvature (with respect to a sphere of same area). Already fixing a value of $c_0=0$ one can retrieve many shapes that are observed in real cells: starting from a sphere ($v=1$) as the internal volume is decreased, prolate ellipsoids transition to dumbell shapes. Than a discontinuous transition leads to biconcave symmetric shapes similar to those observed in the red blood cells. At even smaller volumes, simply concave shapes (like stomatocytes) are the most stable. If one allows spontaneous curvatures to range from negative to positive values, even more complex features can be explained. In particular, prolate shapes dominate for $c_0>2$. At high values of c_0 symmetric dumbbell configurations become unstable and pear-shaped, asymmetric ones appear. Approaching the limit line L^{pear} (the curved solid line at the top of fig. 10a) the dumbbells obtain narrow necks, which are loci of high and equal principal curvatures with opposite signs: past this limit line the neck collapses and the vesicle undergoes a so called budding (the smaller vesicle is ejected from the bigger one). At negative spontaneous curvatures instead oblate shapes dominate and can even lead to the inclusion of spherical cavities in the vesicle for large negative c_0 values. A remarkable transition is observed for oblate vesicles with slightly positive c_0 that decrease in volume: the vesicle self-intersects (this happens crossing the lines SI^{ob}, SI^{sto} from right to left) and acquires a toroidal shape.

46   Helfrich 1973.
47   Seifert/Lipowsky 1990; Seifert/Berndl/Lipowsky 1991.
48   Svetina/Zeks 1983; Svetina/Zeks 1989.

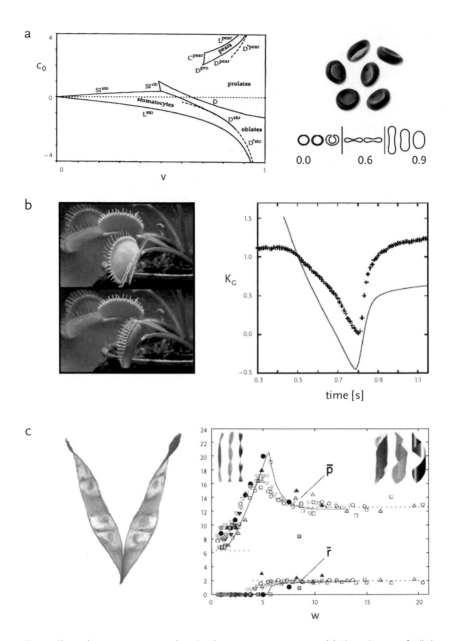

Fig. 10: Shape changes in systems endowed with a spontaneous curvature. (a) Phase diagram of cell shapes. Depending on the cell volume v and its spontaneous curvature c_0 a variety of shape changes is possible: prolate, oblate, concave, biconcave (like the red blood cells), toroidal cell shapes and even cell budding. Seifert/Berndl/Lipowsky 1991. (b) Snapping closure of the Venus flytrap leaves is actively triggered by the plant (slow decrease of Gaussian curvature K_G) but the high speed of the motion (fast increase of K_G) is due to its doubly curved geometry (solid line, theoretical model; markers, experimental measurements). Forterre et al. 2005. (c) Chiral opening of the *Bauhinia Variegata* seed pods results from a competition between stretching and bending dominated behaviors, that eventually depend on the geometry of the pods (its width w) and non-developable target metric/curvature defined by the pod valves' material structure. The pitch ($\bar{p}$) and radius ($\bar{r}$) of the pod valve mark two possible shapes: a twisted helix or a ribbon (solid line, theoretical model; markers, experimental measurements). Armon et al. 2011.

Another example of thin shell with non-developable initial shape (fig. 5i) is the plant known as Venus flytrap. This plant has typical doubly curved leaves that are convex in the open state and concave in the closed state. The morphing between these two states happens through a fast snap that is used to trap a prey. Measuring the leaf Gaussian curvature K_G as a function of time. Forterre and Mahadevan[49] realized that it was not constant during the closure (see fig. 10b): it would first decrease slowly (when the leaf is still convex) and then increase very rapidly (when the leaf becomes concave). As we have seen for the blooming lily, because the leaf is doubly-curved, the curvature inversion (convex-to-concave) can only happen through an intermediate state in which the leaf is stretched. To achieve this, the internal (turgor) pressure of the cells on the external side of the leaf is actively increased until the leaf is close to a critical curvature value. In this stage elastic energy is stored in the stretched leaf. As the turgor pressure is generated by a slower biochemical process K_G decreases slowly. When a prey stimulates the plant "sensors" (cilia on the leaf inner face) only an additional small increase in turgor pressure is needed: the critical threshold is passed, the elastic energy released and the leaf switches to the more stable concave configuration. Moreover, since the Gaussian curvature K_G is tipically higher than in the lily, the curvature inversion happens rapidly through a snap-through buckling. Therefore the overall mechanism ensures fast actuation, which is crucial for its prey-catching effectiveness.

The second scenario (non-developable target shape, fig. 5h) is well depicted by the seed pods of the plant *Bauhinia variegatae*: when the pod dries the two valves split open with a spiralling movement. Cues to understand this movement can be found already in the structure of the valves. As observed by Armon,[50] each valve consist of two fibrous layers oriented roughly at $\pm 45°$ with respect to the pod's longitudinal axis: therefore (recalling how wood actuation works, see section 1.2.2) the two layers shrink uniaxially at perpendicular directions. While each individual layer would stay flat, when they are attached together they develop opposite principal curvatures along the $\pm 45°$ orientation of the fibers. With reference to fig. 5h where the principal directions are along the sheet sides, the valve behaves as a sheet with target metric[51] $a_0 = [0\ 0]$ (where 0 means that no expansion/contraction is allowed due to the presence of fibers at perpendicular directions) and a target double curvature $b_0 = [K_0\ -K_0]$. But such a case is special: the valve locally assumes a saddle-like configuration to satisfy the target curvature with a negative Gaussian curvature, the target metric requires null in-plane strains of the sheet; that is there is a metric incompatibility, where target metric and target curvature cannot be satisfied simultaneously. Therefore the strip will always have a residual stress.

---

49   Forterre et al. 2005.

50   Armon et al. 2011.

51   Target metric and target curvature are equivalent to eigenstrain and eigencurvature. The notation used here was given in section 2.1.

Such metric incompatibility is at the base of a strange behavior: despite having the same intrinsic curvature it was observed that wide strips adopt a ribbon shape (that is a cylindrical envelope shape) whereas narrow strips show a helical one (where the strip twists around its straight centerline). This can be deduced by the plot of fig. 10c where the experimentally measured pitch ($\bar{p}$) is plotted versus radius ($\bar{r}$) for many biological and artificial samples. Which configuration will be acquired depends on the minimum of the total energy. From energy scaling arguments Armon showed that when the strip width w is large the system is stretching-dominated; this implies that to minimize the total energy one has to minimize the stretching energy, where the system satisfies the target metric $a = a_0$. Under this constraint the sheet assumes a curvature different from the target $b \neq b_0$, approximating it in one of the two principal directions while frustrating it (zeroing it) in the other: $b = [\kappa_0 \ 0]$. This corresponds to the ribbon shape. Conversely, when the strip is narrow the sheet is bending dominated, and the energy is minimized for $b = b_0$. Under this constraint, a has to approximate a_0. This corresponds to the helical configuration with twisted straight axis: at the midline the stretching energy is zero, but it increases with w. The switch from one regime to the other happens at a critical width $w_{cr} \approx 2.5 \sqrt{t/\kappa_0}$ where stretching and bending terms are comparable.

A similar case of sheets with a non-developable target shape, is observed in rectangular polymer-metal bilayers activated by an electrical-driven uniform contraction of the polymer layer (fig. 11a). Remarkably Alben[52] observed that bilayers with a large length-to-width (aspect) ratio bend preferentially keeping the short side straight. Because the polymer contracts isotropically, the bilayer assumes a target spherical curvature $b_0 = [\kappa_0 \ \kappa_0]$ (fig. 5g). As we know, such a double curvature implies in-plane stretching, which is an unfavorable mode of deformation. When the polymer contracts, the bilayer first assumes a shallow spherical shape; afterwards it frustrates one of the two curvatures and bends along the other, assuming a cylindrical rolled shape. So why the short-side rolling[53] would be always preferred? Alben realized that the bilayer would not bend into a perfect cylinder, but would preserve narrow regions of double-curvature close to the rolled edges. Since these regions are closer to the target curvature, they lowered the elastic energy of the system. Therefore the system prefers to roll on the short-sides thus maximizing the regions of double-curvature to the long sides.

2.4.3 Sheets with a double eigencurvature confined at the edges
This section on bilayers deals with the morphing of edge activated bilayers (fig. 5j) where a spontaneous double curvature $b_0 = [\kappa_0 \ \kappa_0]$ appears at the edges. In a set of experiments on swelling polymer bilayers of variable rectangular shape, Stoychev[54] observed a variety of

---

52 Alben/Balakrisnan/Smela 2011.
53 In short-side rolling the short side stays flat and the long ones bends.
54 Stoychev et al. 2012.

rolling scenarios (fig. 11b): long side rolling in case of large aspect ratio bilayers; diagonal and all-side rolling in case of aspect ratios close to 1. Strikingly enough, short side rolling was not observed at all: Stoychev's was in open contradiction with the experiments of Alben. The main reason for the different behavior is due to the different activation. The polymer bilayers synthesized by Stoychev showed a swelling (active) layer attached to the substrate and a passive (non-swelling) layer on top. Since the active layer is confined between the substrate and the passive layer, water can diffuse in it only through the lateral faces (edges), and consequently swelling will occur first close to the edge and then reach more internal portions of the bilayer with time. Because swelling is confined to the edges, the bilayer does not transit through the shallow spherical shape. Therefore rolling starts at the corners first (where the diffusion depth is maximum) creating conical shapes; then two scenarios are possible. Long-side rolling is energetically favored for high aspect ratio because the costly in-plane stresses are relaxed (by bending) in a larger portion (the long edge). When instead the aspect ratio of the bilayer is about one, the bilayer can progress towards diagonal rolling (if rolling was initiated at two opposite corners) or all-side rolling (if it started at adjacent ones).

In a subsequent paper, Stoychev[55] investigated the folding of polymer bilayers with more general shapes, such as ellipses or star shapes with a round contour (fig. 11c). In the early stages of swelling Stoychev observed the emergence of partially rolled ripples at the rounded edges, bestowing them with a characteristic "beer cap" contour. Moreover the number of ripples would decrease as the diffusion front travels deeper towards the center. Such edge rippling is in agreement with the edge growth in single sheets but, since the sheet is a bilayer, the edges preferentially bend in one out-of-plane direction, that is, the expected edge rippling will not be symmetric. At the same time it contradicts the formation of edge tubes observed in rectangular bilayers. In reality both processes occur, although at different time scales: the initial ripples would fuse together when the angle between them exceeds 130° (this happens always for edges that are not too convex) forming the familiar tubular edge. Once long rolls are formed, they make the bilayer more rigid. As a result, after the edge tube formation bilayers can only fold over the lines connecting weak points between tubular edges. These movements resemble the folding operations in artistic origami, in which a flat sheet of paper is folded along specific lines to obtain a three-dimensional figure. It is remarkable that such a simple system characterized by isotropic mechanical and swelling properties undergoes complex multi-step actuation: again, the eventual folded shape depends from the previous stages acquired during the progression of the morphing process.

---

55   Stoychev et al. 2013.

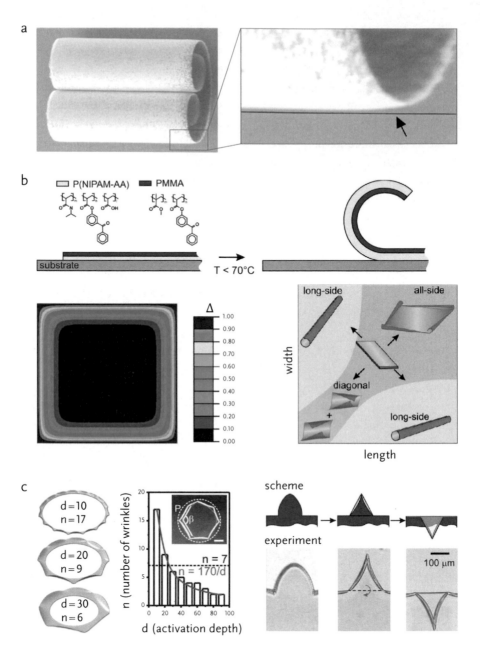

Fig. 11: Shape selection and evolution in bilayer systems with a spontaneous curvature subjected to edge effects. (a) Short side rolling is the preferred deformation mode in narrow bilayer strips in case of double eigencurvature: in this way the long edges maximize the extent of doubly curved edge portions. Alben/ Balakrisnan/Smela 2011. (b) Scheme of bilayer composition (top), edge activation Δ (colored square) and phase diagram of rolling modes when the eigencurvature is activated at the edges first: in this case long side rolling is preferred because a doubly curved state of the edges is prevented from the initial adhesion to a substrate. Stoychev et al. 2012. (c) Multistep asymmetric ripple formation and flap folding in swelling bilayers with complex star shapes: ripples form at the edges and (depending on their incident angles) fuse to rolls. When the edge is rolled the star arms are stiffened leaving only origami-like foldings as possible deformation. Stoychev et al. 2013.

2.5 Self-folding origami

As a concluding remark we would like to at least mention a *separate* group of self-folding systems which is recently gaining more and more attention: the artificial self-folding origami. These systems are designed to experience extremely large eigenstrains and eigencurvatures at very narrow locations in the sheet: in fact a typical crease pattern appears, whereas the rest of the sheet is practically undeformed. This aspect makes such systems extremely convenient to be studied because their deformation depends only on the extent and time order of rigid body rotations at the creases. Bound to this is their success in robotic applications: to give an example, in a recent paper Felton[56] demonstrated a prototype of robot that is able to fold itself from a flat creased plate, hold its own weight and even walk.

3. Conclusion

Although the present work is far from being exhaustive, we hope we have shed light on the underlying mechanisms that govern spontaneous shape change in thin systems. Despite the governing equations of plates subjected to eigenstrain being known since one century[57], this branch of research is still quite active. This is due, in our opinion, to three main factors. First of all, these equations are scaleless: as we have seen in our examples many morphing processes could be explained regardless of their actual size, and therefore they can be applied to disparate systems. Second, the variety of patterns that can be predicted derives from the intrinsic nonlinearity of the elastic strain formulation[58]: this makes the theory appealing to explain experimental observations in which little is known about the material properties (due to difficulties bound to the characterization methods or samples accessibility/availability). And third, especially relevant for theoretical research and simulation, is that there is a virtually infinite panorama of possible combinations of eigenstrains/eigencurvatures distributions, plate shapes and boundary conditions. In this perspective, the present paper is especially useful: presenting the folding of many different thin structures under the unifying concept of eigenstrain (and as consequence thereof), it enables to easily locate where research efforts have been focused until now. As a consequence, the paper calls for an exploration of less explored eigenstrain distributions: in turn, this will foster a broader and deeper understanding of self-folding systems and lead to new exciting engineering applications.

---

56   Felton et al. 2014.
57   Föppl 1907; von Kármán 1910.
58   The reader can verify that in all the examples the plate material has always been considered linear elastic.

Bibliography

Alben, Silas / Balakrisnan, Bavani / Smela, Elisabeth (2011): *Edge Effects Determine the Direction of Bilayer Bending*. In: Nano Letters, vol. 11, no. 6, pp. 2280–2285.

Armon, Shahaf / Efrati, Efi / Kupferman, Raz / Sharon, Eran (2011): *Geometry and Mechanics in the Opening of Chiral Seed Pods*. In: Science, vol. 333, no. 6050, pp. 1726–1730.

Audoly, Basile / Boudaoud, Arezki (2003): *Self-similar structures near boundaries in strained systems*. In: Physical Review Letters, vol. 91, no. 8, 086105.

Audoly, Basile / Boudaoud, Arezki (2008a): *Buckling of a stiff film bound to a compliant substrate – Part I: Formulation, linear stability of cylindrical patterns, secondary bifurcations*. In: Journal of the Mechanics and Physics of Solids, vol. 56, no. 7, pp. 2401–2421.

Audoly, Basile / Boudaoud, Arezki (2008b): *Buckling of a stiff film bound to a compliant substrate – Part II: A global scenario for the formation of herringbone pattern*. In: Journal of the Mechanics and Physics of Solids, vol. 56, no. 7, pp. 2422–2443.

Audoly, Basile / Boudaoud, Arezki (2008c): *Buckling of a stiff film bound to a compliant substrate – Part III: Herringbone solutions at large buckling parameter*. In: Journal of the Mechanics and Physics of Solids, vol. 56, no. 7, pp. 2444–2458.

Audoly, Basile / Pomeau, Yves (2010): *Elasticity and geometry: from hair curls to the nonlinear response of shells*. Oxford: Oxford University Press.

Bowden, Ned / Brittain, Scott / Evans, Anthony G. / Hutchinson, John W. / Whitesides, George M. (1998): *Spontaneous formation of ordered structures in thin films of metals supported on an elastomeric polymer*. In: Nature, vol. 393, no. 6681, pp. 146–149.

Brau, Fabian / Damman, Pascal / Diamant, Haim / Witten, Thomas (2013): *Wrinkle to fold transition: influence of the substrate response*. In: Soft Matter, vol. 9, no. 34, pp. 8177–8186.

Breid, Derek / Crosby, Alfred J. (2013): *Curvature-controlled wrinkle morphologies*. In: Soft Matter, vol. 9, no. 13, pp. 3624–3630.

Cai, Shengqiang / Breid, D. / Crosby, A. J. / Suo, Zhigang / Hutchinson, John W. (2011): *Periodic patterns and energy states of buckled films on compliant substrates*. In: Journal of the Mechanics and Physics of Solids, vol. 59, no. 5, pp. 1094–1114.

Callister, William D. (2007): *Materials science and engineering: an introduction*. New York: John Wiley & Sons.

Cao, Guoxin / Chen, Xi / Li, Chaorong / Li, Ailing / Cao, Zexian (2008): *Self-assembled triangular and labyrinth buckling patterns of thin films on spherical substrates*. In: Physical Review Letters, vol. 100, no. 3, 036102.

Deleuze, Gilles (1993): *The fold: Leibniz and the Baroque*. trans. Conley, Tom. London: The Athlone Press.

Dias, Marcelo A. / Hanna, James A. / Santangelo, Christian D. (2011): *Programmed buckling by controlled lateral swelling in a thin elastic sheet*. In: Physical Review E, vol. 84, no. 3, 036603.

Dunlop, John W. C. (2016): *The physics of shape changes in biology*, Habilitation, Universität Potsdam.

Efrati, Efi / Sharon, Eran / Kupferman, Raz (2009a): *Elastic theory of unconstrained non-Euclidean plates*. In: Journal of the Mechanics and Physics of Solids, vol. 57, no. 4, pp. 762–775.

Efrati, Efi / Sharon Eran / Kupferman, Raz (2009b): *Buckling transition and boundary layer in non-Euclidean plates*. In: Physical Review E, vol. 80, no. 1, 016602.

Efrati, Efi / Sharon Eran / Kupferman, Raz (2011): *Hyperbolic non-Euclidean elastic strips and almost minimal surfaces*. In: Physical Review E, vol. 83, no. 4, 046602.

Efrati, Efi / Sharon Eran / Kupferman, Raz (2013): *The metric description of elasticity in residually stressed soft materials*. In: Soft Matter, vol. 9, no. 34, pp. 8187–8197.

Elbaum, Rivka / Zaltzman, Liron / Burgert, Ingo / Fratzl, Peter (2007): *The role of wheat awns in the seed dispersal unit*. In: Science, vol. 316, no. 5826, pp. 884–886.

Euler, Leonhard (1953): *Opera Omnia*. Zurich: Orell Füssli Verlag.

Felton, Samuel M. / Tolley, Michael / Demaine, Erik / Rus, Daniela / Wood, Robert (2014): *A method for building self-folding machines*. In: Science, vol. 345, no. 6197, pp. 644–646.

Föppl, August (1907): *Vorlesungen über technische Mechanik*. Leipzig: B. G. Teubner.

Forterre, Yoel / Skotheim, Jan M. / Dumais, Jacques / Mahadevan, Lakshminarayanan (2005): *How the Venus flytrap snaps*. In: Nature, vol. 433, no. 7024, pp. 421–425.

Fratzl, Peter / Elbaum, Rivka / Burgert, Ingo (2008): *Cellulose fibrils direct plant organ movements*. In: Faraday Discussions, vol. 139, pp. 275–282.

Gambhir, Murari Lal (2004): *Stability analysis and design of structures*. Berlin, Heidelberg, New York: Springer-Verlag.

Gauss, Karl Friedrich (1902): *General investigations of curved surfaces of 1827 and 1825*. Princeton: The Princeton University Library.

Genzer, Jan / Groenewold, Jan (2006): *Soft matter with hard skin: From skin wrinkles to templating and material characterization*. In: Soft Matter, vol. 2, no. 4, pp. 310–323.

Harrington, Matthew J. / Razghandi, Khashayar / Ditsch, Friedrich / Guiducci, Lorenzo / Rueggeberg, Markus / Dunlop, John W. C. / Fratzl, Peter / Neinhuis, Christoph / Burgert, Ingo (2011): *Origami-like unfolding of hydro-actuated ice plant seed capsules*. In: Nature Communications, vol. 2, no. 337, pp. 1–7.

Helfrich, Wolfgang (1973): *Elastic Properties of Lipid Bilayers – Theory and Possible Experiments*. In: A Journal of Biosciences: Zeitschrift für Naturforschung C, vol. 28, no. 11, pp. 693–703.

Huang, Zhiyong Y. / Hong, Wei / Suo, Zhigang (2005): *Nonlinear analyses of wrinkles in a film bonded to a compliant substrate*. In: Journal of the Mechanics and Physics of Solids, vol. 53, no. 9, pp. 2101–2118.

Ivanov, Valentin I. / Trubetskov Michael K. (1995): *Handbook of Conformal Mapping with Computer-Aided Visualization*. Boca Raton: CRC Press.

Kim, Jungwook / Hanna, James A. / Byun, Myunghwan / Santangelo, Christian D. / Hayward, Ryan C. (2012): *Designing Responsive Buckled Surfaces by Halftone Gel Lithography*. In: Science, vol. 335, no. 6073, pp. 1201–1205.

Klein, Yael / Efrati, Efi / Sharon, Eran (2007): *Shaping of elastic sheets by prescription of non-Euclidean metrics*. In: Science, vol. 315, no. 5815, pp. 1116–1120.

Li, Bo / Jia, Fei / Cao, Yan-Ping / Feng, Xi-Qiao / Gao, Huajian (2011): *Surface Wrinkling Patterns on a Core-Shell Soft Sphere*. In: Physical Review Letters, vol. 106, no. 23, 234301.

Liang, Haiyi / Mahadevan, Lakshminarayanan (2009): *The shape of a long leaf*. In: Proceedings of the National Academy of Sciences of the United States of America, vol. 106, no. 52, pp. 22049–22054.

Liang, Haiyi / Mahadevan, Lakshminarayanan (2011): *Growth, geometry, and mechanics of a blooming lily*. In: Proceedings of the National Academy of Sciences of the United States of America, vol. 108, no. 14, pp. 5516–5521.

Lin, Pei-Chun / Yang, Shu (2007): *Spontaneous formation of one-dimensional ripples in transit to highly ordered two-dimensional herringbone structures through sequential and unequal biaxial mechanical stretching*. In: Applied Physics Letters, vol. 90, no. 24, 2419031.

von Kármán, Theodore (1910): *Festigkeitsproblem im Maschinenbau*. In: Enzyklopädie der mathematischen Wissenschaften, vol. IV, pp. 311–385.

Pocivavsek, Luka / Dellsy, Robert / Kern, Andrew / Johnson, Sebastián / Lin, Binhua / Lee, Ka Yee C. / Cerda, Enrique (2008): *Stress and fold localization in thin elastic membranes*. In: Science, vol. 320, no. 5878, pp. 912–916.

Pressley, Andrew (2009): *Elementary differential geometry*. London: Springer.

Seifert, Udo / Berndl, Karin / Lipowsky, Reinhard (1991): *Shape Transformations of Vesicles – Phase-Diagram for Spontaneous-Curvature and Bilayer-Coupling Models*. In: Physical Review A, vol. 44, no. 2, pp. 1182 – 1202.

Seifert, Udo / Lipowsky, Reinhard (1990): *Adhesion of Vesicles*. In: Physical Review A, vol. 42, no. 8, pp. 4768 – 4771.

Sharon, Eran / Roman, Benoît / Marder, Michael / Shin, Gyu-Seung / Swinney, Harry L. (2002): *Mechanics: Buckling cascades in free sheets – Wavy leaves may not depend only on their genes to make their edges crinkle*. In: Nature, vol. 419, no. 6907, pp. 579 – 579.

Stoop, Norbert / Lagrange, Romain / Terwagne, Denis / Reis, Pedro M. / Dunkel, Jörn (2015): *Curvature-induced symmetry breaking determines elastic surface patterns*. In: Nature Materials, vol. 14, no. 3, pp. 337 – 342.

Stoychev, Georgi / Zakharchenko, Svetlana / Turcaud, Sebastien / Dunlop, John W. C. / Ionov, Leonid (2012): *Shape-Programmed Folding of Stimuli-Responsive Polymer Bilayers*. In: ACS Nano, vol. 6, no. 5, pp. 3925 – 3934.

Stoychev, Georgi / Turcaud, Sebastien / Dunlop, John W. C. / Ionov, Leonid (2013): *Hierarchical Multi-Step Folding of Polymer Bilayers*. In: Advanced Functional Materials, vol. 23, no. 18, pp. 2295 – 2300.

Svetina, Saša / Zeks, Boštjan (1983): *Bilayer Couple Hypothesis of Red-Cell Shape Transformations and Osmotic Hemolysis*. In: Biomedica Biochimica Acta, vol. 42, no. 11 – 12, pp. S86 – S90.

Svetina, Saša / Zeks, Boštjan (1989): *Membrane Bending Energy and Shape Determination of Phospholipid-Vesicles and Red Blood-Cells*. In: European Biophysics Journal with Biophysics Letters, vol. 17, no. 2, pp. 101 – 111.

Thompson, D'Arcy Wentworth (1953): *On growth and form*. Cambridge: University Press.

Lorenzo Guiducci
Email: *lorenzo.guiducci@mpikg.mpg.de*

Image Knowledge Gestaltung. An Interdisciplinary Laboratory.
Cluster of Excellence Humboldt-Universität zu Berlin.
Sophienstrasse 22a, 10178 Berlin, Germany.

Department of Biomaterials, Max Planck Institute of Colloids and Interfaces.
14476 Potsdam, Germany.

John W. C. Dunlop; Peter Fratzl
Department of Biomaterials, Max Planck Institute of Colloids and Interfaces.
14476 Potsdam, Germany.

Mohammad Fardin Gholami, Nikolai Severin, Jürgen P. Rabe

Folding of Graphene and Other Two-dimensional Materials

1. Introduction

1.1 Folding

The concept of the fold, from a physical standpoint, has been a subject of investigation, spanning from folding of proteins[1] and synthetic polymers[2] to packaging engineering[3]. Folding of fabrics and thin sheets of material has played an important part in advances made in tool production (folding of solar panels, for example, is crucial for efficiency in space utilization aboard satellites).

Folding allows for the transformation of a two-dimensional (2D) material into a three-dimensional (3D) material that exhibits a more complex geometry. The importance of folding can be observed in nature from the structure of plant leaves and animal tissues to protein folding; it is the most important feature of their functionality.

Nature takes a bottom-up approach to constructing functional foldable components, such as plant leaves and animal tissues, by assembling together smaller functional units, thereby giving rise to progressively more complex systems. Against this backdrop, the use of atomically thin sheets such as *graphene*, which are able to fold themselves into various geometries, is clearly interesting.

---

1   Dill/McCallum 2012.
2   Shu et al. 2001; Zhuang et al. 2008; Schmidt et al. 2011.
3   Fei/Debnath 2013.

Before proceeding to discuss folding at nanoscale levels, let us first address folding and bending of materials at macroscale levels. Folding and bending are natural processes present, for example, in plant growth. At the macroscale level, in which the thickness of a sheet of material is greater than a micrometer, high enough strain across the bulk results in bending and folding.

In nature, forces, which cause bending or folding, could result from *differential growth* imposed by various external factors, such as changes in humidity or thermal expansions.[4] Differential growth can occur when layers have different mechanical properties, leading to differential stress experienced at each layer, a differentiation that results in bending and consequently folding.

Furthermore, folding and self-folding of materials could be utilized towards nanoactuation, stimuli responsive folding, and wrapping. On the nanoscale, e.g., in graphene (fig. 1), where the thickness of a sheet of material is on the order of an atomic or molecular diameter, the folding is possible with the tip of a scanning force microscope (SFM).[5]

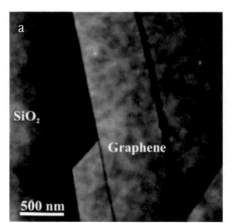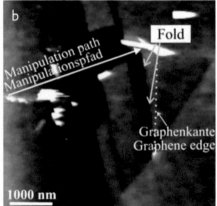

Fig. 1: Scanning Force Microscopy (SFM) height images display the topography of single graphene layers on a silicon oxide (SiO_2) surface before (a) and after (b) manipulation with the SFM. Higher regions appear brighter; a white arrow indicates the path the SFM probe was moved during manipulation; the dotted line shows the graphene edge after folding.

Also a variation of the environmental conditions and *chemical patterning*[6] may be employed. The possibility of nanoscale folding in materials promises the introduction of novel properties into matter as well as new applications for resulting products.

---

4   Harrington et al. 2011.
5   Eilers/Rabe 2009.
6   By chemical patterning we mean the controlled variation of the chemistry within a single sheet of graphene or graphene oxide.

In the following, section 1.2 is devoted to describing properties of graphene and graphene oxide. Section 1.3 then continues with an overview of recent studies related to their folding. In section 2 we describe folding of 2D nanoscale sheets and the effects of defect lines on possible bending and folding. Section 3 explains the impact of folding on the electronic properties of graphene. Section 4 presents experiments and current research on folding of graphene and graphene oxides, focusing on their possible applications for nanotechnology, especially in nanopackaging.

1.2 Graphene and graphene oxide

Graphene is a single-atom thick sheet of carbon atoms arranged in a two-dimensional network of ring structures (a honeycomb lattice).[7] Andre Geim et al. were the first to isolate graphene through exfoliation of a naturally occurring graphite crystal using common adhesive tapes.[8] Subsequent studies demonstrated graphene has many promising properties such as high electrical conductivity at room temperature[9] and high thermal conductivity,[10] and on the same time high transparency[11] moreover, it exhibits a low bending stiffness,[12] and nevertheless a very high amount of force (per unit area) is required for stretching (also known as high Young's modulus).[13]

In addition, graphene has been proven to be impermeable to small molecules.[14] This impermeability is attributed to its dense delocalized cloud of electrons, effectively blocking the ring structure with a repelling field.[15] However, as with any other crystalline structure, defects and incomplete segments may exist in random locations throughout the graphene sheet.

Simulation studies and microscopic methods have already demonstrated that graphene, as a 2D sheet of atomic carbon cores, can undergo 3D deformations to minimize its energy.[16] This was attributed to structural defects.[17] Structural defects thus become a crucial element for 3D deformations on the nanoscale.

---

7   Novoselov et al. 2004; Geim/Novoselov 2007; Meyer et al. 2007.
8   Novoselov et al. 2004; Wissler 2006.
9   Le/Woods 2012.
10   Calizo et al. 2008.
11   Nair et al. 2008; Cai et al. 2009.
12   See footnote 8.
13   Lee et al. 2008; Castro Neto et al. 2009; Le/Woods 2012.
14   Bunch et al. 2008; Lange et al. 2011; Li et al. 2013.
15   Sreeprasad/Berry 2013.
16   Roy et al. 1998; Zhang et al. 2010; Wang et al. 2013; Zhang/Li/Gao 2014.
17   Zhang et al. 2010.

Consequently the discovery of folds and wrinkles in graphene (2D to 3D transformation) has attracted much interest for their possible use in various applications, involving either graphene itself or one of its most important chemical precursors for its chemical synthesis, *graphene oxide* (GO).

Graphene oxide is a highly oxidized version of graphene.[18] It was initially produced by Benjamin Brodie in 1859, while trying to measure the atomic weight of graphite.[19] The two substances differ in that graphene is an electron conductor, while graphene oxide is an insulator. This is attributed to defects and structural disruptions caused by the presence of oxygen containing groups within GO.[20]

Researchers favor it for its potential for chemical modification; moreover, GO is easily processed in large quantities, because it can be synthesized by oxidization of graphite crystals.[21] Graphene oxide is highly dispersible. It forms a suspended mixture – *colloidal suspension* – in polar solvents such as water, due to the presence of oxygen containing groups. A colloidal suspension of graphene oxide was shown to be a key requirement in the transformation of the GO sheets from 2D to 3D structures. In addition, colloidal suspension of GO is an efficient and up-scalable (large scale production) way for the chemical reduction of GO.[22]

The self-assembly of graphene oxide sheets around nanomaterials has recently been investigated; it was found to be governed for the most part by electrostatic interactions.[23] Graphene oxide has its own spectrum of applications as in super capacitors,[24] membrane technology,[25] batteries,[26] and sensors[27] or as a part of composite materials.[28]

1.3 Folded graphene

Folded graphene exhibits further interesting properties, such as alteration of electron mobility (conductivity) at folded regions.[29] The folding of graphene sheets has been

18   Hofmann/Holst 1939.
19   Brodie 1859.
20   Dikin et al. 2007; Dreyer et al. 2010.
21   Hummers/Offeman 1958; Chen et al. 2013.
22   Lee et al. 2010; Zhou 2011.
23   Cassagneau/Fendler 1999; Yang et al. 2010; Zhu/Chen/Liu 2011.
24   Stoller et al. 2008; Chen et al. 2010.
25   Dreyer et al. 2010; Li et al. 2013; Joshi et al. 2014.
26   Tao et al. 2011; Wang et al. 2012.
27   Cheng-Long et al. 2011; Borini et al. 2013.
28   Wu et al. 2015.
29   Prada/San-Jose/Brey 2010.

observed to result in the formation of graphene rolls or *nanotubes*. Moreover, graphene and graphene oxide have each been the subject of many studies regarding their self-transformation from 2D to 3D. Examples of such studies can be found in Nano-origami,[30] Kirigami,[31] biological applications[32] and even wrapping of bacteria.[33] As a form of folded graphene, carbon nanotubes have been demonstrated useful in fast water transport at nanoscale levels.[34]

In order to transform graphene into desired geometries relevant to new applications, many methods have been introduced for shaping or patterning.[35] However, some exciting new applications, such as nanocages devised for storage purposes, have only been realized in theoretical simulations.[36] The main limitation in experimentally realizing them appears to be a lack of control at the nanoscale, while full control over the different forces operating at the nanoscale is required in order to achieve ease of transformation from 2D into a stable 3D configuration.[37]

2. Mechanics of folding in graphene

In section 2.1 we describe the folding mechanism of atomically thin graphene sheets in comparison with thick multilayered materials. In section 2.2 we follow with a discussion of the influence of structural defects on folding and present related studies.

2.1 Folding in 2D nanoscale sheets

One may consider folding as an ultimate form of bending where a sheet of material overlaps but does not intersect itself while it remains in a stable state. To understand folding at the nanoscale, bending must first be defined and understood. Recent research has shown that the process of bending in single-atom thick or multilayered graphene sheets can be modeled using finite *deformation beam theory*.[38] This suggests bending of a multilayered thick sheet of graphene is not considerably different from that of a sheet of paper.

30   Patra/Wang/Král 2009.
31   Blees et al. 2015.
32   Chung et al. 2013; Kostarelos/Novoselov 2014.
33   Liu et al. 2011; Akhavan/Ghaderi/Esfandiar 2011; Mohanty et al. 2011.
34   Ma et al. 2015.
35   Feng et al. 2012; Wang et al. 2013; Chen et al. 2014.
36   Leenaerts/Partoens/Peeters 2008; Zhang/Zeng/Wang 2013.
37   Patra/Wang/Král 2009; Talapin et al. 2007.
38   Tarsicio/Cristian/Augusto 2002; Xianhong et al. 2013.

As a simple example, cylindrical bending along a given longitudinal direction in a sheet of graphene is conceptually the main process responsible for the formation of single wall carbon nanotubes. It is similar to the rolling of a sheet of paper. Therefore, the resistance of a material to bend – that is, its *bending modulus* – for graphene would be crucial for understanding and utilizing bending mechanics.[39]

Empirical potentials [40] and *ab initio* calculations [41] have been previously used to predict the bending modulus of graphene. Calculations of the bending stiffness of graphene are not simple, since the thickness of a single-layer graphene, which is determined by quantum mechanical methods, cannot be defined in the same as on the macroscopic scale. Introducing bending stiffness narrows further the analogy with a sheet of paper.

A bending modulus can be introduced, assuming graphene to be a homogeneous sheet with a certain thickness. Yet introducing a bending modulus is rather a mathematical trick and does not imply graphene to be a thin elastic plate, since the bending modulus for an elastic thin plate depends on cube of thickness (that is to the third power).[42]

According to Kirchhoff-Love theory of plates, the strain and stress along the bulk of a thin homogeneous elastic plate varies linearly with thickness. Lu Qiang et al.[43] used a first-generation *Brenner potential*[44] to derive a simple analytical form for the bending modulus of single-atom thick graphene. The value of the bending stiffness was found to be 0.133 nN nm, while *ab initio* calculations have shown it to be 0.238 nN nm.[45]

It is clear then that values obtained through *ab initio* methods predict a stiffer graphene. Furthermore, this suggests that the empirical model does not explain the bending stiffness of the single-atom thick graphene sheet. Qiang et al. derived a formula for the bending stiffness of single-atom thick graphene based on an empirical potential for solid-state carbon atoms. It considers the bending stiffness of graphene not only being dependent on bond angle effects but also on the out-of-plane twisting angle between carbon-carbon bonds, known as torsion angles or *dihedral angles* (see fig. 2).[46] They calculated a bending stiffness of 0.225 nN nm, very close to the *ab initio* calculations. The dihedral angle effect

39   Qiang/Marino/Rui 2009.
40   Tersoff 1988; Tu/Ou-Yang 2002; Marino/Belytschko 2004; Huang/Wu/Hwang 2006.
41   Sánchez-Portal et al. 1999; Kudin/Scuseria/Yakobson 2001.
42   Timoshenko/Woinowsky 1987.
43   Qiang/Marino/Rui 2009.
44   This model is a function for calculating the potential energy of covalent bonds and the interatomic force. The total potential energy of a system is the sum of nearest-neighbor pair interactions which depend not only on the distance between atoms but also on their local atomic environment.
45   Kudin/Scuseria/Yakobson 2001.
46   Qiang/Marino/Rui 2009.

calls for further studies involving bending energies of graphene, as its contribution to bending stiffness of single-atom thick graphene is significant.

a

b

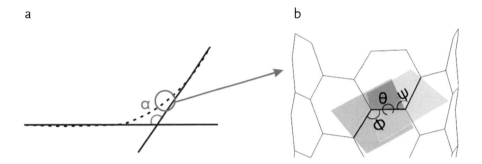

Fig. 2: (a) Graphene sheet bent to an angle α. Orange circle denotes area with bent segment, as detailed in (b). (b) Bent graphene honeycomb structure displaying the carbon-carbon bonds. Each dihedral angle θ (out of plane twisting of C-C bonds) is defined by three bonds (4 carbon atoms) connecting atoms (red lines). The dihedral angle determines the bending. Two neighboring carbon-carbon bond angles are denoted by Φ and Ψ, given by 3 carbon atoms, which define each plane (purple and green, respectively).

In the case of single-atom thick 2D sheets, such as graphene and to a good approximation also single layer graphene oxide, an external force can facilitate the bending process.[47] Researchers have shown that the self-folding of a single-layer graphene sheet can be modeled using similar considerations for calculating load-carrying and deflection characteristics of beams (deformation beam theory).[48]

Using beam theory, it was possible to predict the shape and the critical length of self-folding in graphene. Moreover, *molecular dynamics* (MD) confirmed this theoretical model.[49] A critical length is required for the process of folding in graphene to be autonomous. To visualize such a critical length, let us first define what we mean by a fold in a single-atom thick sheet of graphene.

---

47   Chen et al. 2014; Guo et al. 2011.

48   Tarsicio/Cristian/Augusto 2002; Xianhong et al. 2013.

49   Molecular dynamics is a simulation method usually carried out to simulate physical movements of atoms and molecules governed by various forces such as Van der Waals forces or Coulombs forces.

At its simplest, a graphene fold is defined to be two overlapping sheets of graphene with an arc bending out-of-plane between them (fig. 3a).

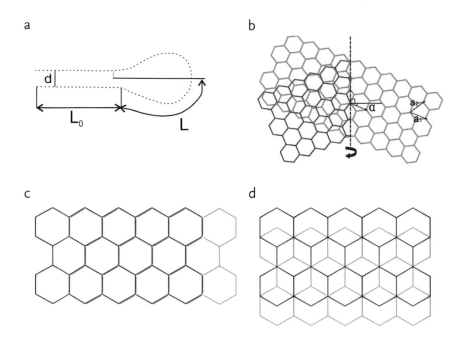

Fig. 3: Conceptual visualizations of (a) profile of a hairpin folded graphene sheet with its denoted regions, where d is the bilayer distance, L_0 and L are the lengths of the flat and half arc (hairpin) regions, respectively. (b) Schematics of a folding graphene sheet with a specific folding angle α. (c) AA stacking of the graphene sheets. (d) AB stacking of the graphene sheets.

The graphene fold involves two flat regions with an arched region in between. Flat regions can adhere to each other due to *van der Waals interactions* forces within an interlayer distance of *d*. These forces are weak attractive forces between two polarizable entities that do not stem from chemical bonding. The arched region is due to the graphene carbon-carbon bonding flexural hindrance.

Consequently, the formation of arched regions requires a certain amount of energy. For a self-folding process to take place, it is important that the adhesion energy between the flat overlapping regions – $E_{adhesion}$ – is greater than the bending energy – $E_{bending}$. Therefore, the length of the overlapping graphene sheet (fig. 3a) directly influences $E_{adhesion}$.

There exists a minimum length of the overlapping graphene segment below which $E_{adhesion}$ < $E_{bending}$, consequently favoring unfolding.

Mohammad Fardin Gholami, Nikolai Severin, Jürgen P. Rabe

To further quantify the total energy required for self-folding the entire length of the fold is needed. The entire length of the fold is:

$$L_{fold} = 2\,(L_0 + L)$$

where L_0 and L stand for the length of the flat region and the length of the half arc region, respectively (see fig. 3a). Meng Xianhong et al. calculate the total energy for folded graphene as follows:

$$E_{fold} = E_{bending} + E_{adhesion}$$

where $E_{adhesion} = -\gamma L_0$, with γ and L_0 standing for the binding energy per unit area of graphene and the length of the flat region, respectively. For self-folding to be possible E_{fold} must be negative. This can only happen when the overlapping length, L_0, is large enough and consequently $E_{adhesion}$ is higher in absolute value.[50] For any interlayer distance, d, between overlapping flat regions, E_{fold} can be calculated with respect to L, the half-length of the arched region. Therefore, the self-folding energy can be evaluated for different sheet geometries by:

$$(1) \qquad E_{fold} = E_{bending} - \frac{\gamma}{2}\,(L_{fold} - 2L)$$

Utilizing Eqn. (1), Xianhong et al. carried out MD simulations to investigate self-folding of graphene. MD simulations were carried out assuming a constant temperature of o Kelvin (K) to prevent atomic vibrations in agreement with theoretical models. To initiate the self-folding process an external force was applied in order to bring the graphene sheet ends closer together. Then energy minimization of the system was initiated. This process was repeated for graphene sheets of different lengths so as to identify metastable and stable self-folding. Using this methodology, Xianhong et al. found a critical length of graphene sheets for stable self-folding of $L_{fold} = 11$ nm. They classified the critical length of graphene required for self-folding into three regimes.

1. For graphene lengths L_{fold} below 6.5 nm, self-folding was not possible, and only a 2D configuration was stable.
2. For L_{fold} between 6.5 nm and 10.5 nm, self-folding was metastable, where both 2D and 3D configurations were stable.
3. For L_{fold} longer than 10.5 nm, self-folding or 3D configuration of the graphene sheet was stable.

---

50 Tarsicio/Cristian/Augusto 2002.

Jiong Zhang et al. demonstrated that ultrasonication of graphene sheets in solution at 200 W for ten minutes, leads to the formation of folds in the 2D structure.[51] This implies suspended graphene sheets can fold under mechanical stimuli. They demonstrated that using Transmission Electron Microscopy (TEM) and computer based simulations it is possible to determine the length of the overlapping region and the shape of the curved edge. Based on simulations, the minimum length of the overlapping region should be at least 1.68 nm to form a stable fold. TEM studies of 100 graphene folds revealed that a given fold had either a 0 degree (armchair) or a 30 degree (zigzag) edge (fig. 4).

Zigzag

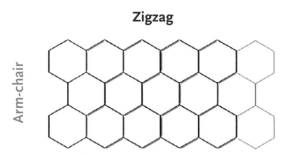

Fig. 4: 0 degree (armchair) or 30 degree (zigzag) fold edges.

To understand their results better, Zhang et al. used further simulations to minimize the energy of graphene folding at different fold edge angles. These simulations showed two prominent fold edge energy minima at 0 and 30 degrees, implying the energy of the folded graphene could be explained through lattice registry effects. Moreover, overlapping regions of graphene folds with AA and AB stacking (fig. 3c and 3d) require the highest and lowest energy states respectively. They suggested their results could be used for the design of graphene-based nanodevices.

These results agree with previous MD simulations by Niladri Patra et al., in which they demonstrated guided folding of graphene sheets.[52] Patra et al. further studied the transformation of 2D graphene sheets into 3D structures at locations where a water droplet was deposited. They observed that, irrespective of the shape of the graphene sheet, graphene could enwrap the water nanodroplets at T = 300 K (room temperature). Such enveloping causes the sheets to bend from their natural 2D configuration into a configuration maintained for a long period of time, even though it is not a minimum energy (metastable) 3D structure.

---

51   Wang et al. 2013.
52   Patra / Wang / Král 2009.

They demonstrated further that, by increasing the temperature of the system from 300 K to 400 K, the shape fluctuations of water nanodroplets increases. Eventually van der Waals attraction between the graphene sheets causes the water droplets to get ejected. Patra et al. calculated the flexural rigidity of the graphene sheets and compared it to theoretically derived values. Their figure was 0.194 nN nm, which is very close to previously calculated values of 0.238 nN nm, 0.11 nN nm, and 0.225 nN nm from other studies.[53]

They therefore considered their simulations to be reasonably close to the possible experimental values. Consequently, the results of MD simulations carried out by Patra et al. imply that nanodroplets activate and guide folding of graphene sheets of complex shapes, similar to chaperones fold proteins.[54] These studies not only demonstrate the possibility of controlling the folding process using water nanodroplets and temperature changes. The latter implies possible applications of water nanodroplets packaging with graphene sheets.

Luca Ortolani et al. investigated the structure of graphene grown by *chemical vapor deposition* (CVD) on copper substrates by transferring graphene onto a TEM grid.[55] They used *high-resolution transmission electron microscopy* (HRTEM) to study graphene sheets at folded regions. In order to map out the curvature at the folds, compressions, observed in TEM images of the graphene lattice, were subject to *geometric phase analysis* (GPA).[56] Data obtained from GPA analysis then helped them to reconstruct variations in height and local curvature of edges of graphene folds within a spatial resolution of 0.5 nm. Ortolani et al. propose the analysis of apparent strains in HRTEM images of general 2D sheets as a way to achieve sub-nanometer topographic information experimentally.

2.2 Defect guided wrinkling or folding in graphene

In the structure of an atomically thin sheet of graphene, there may be regions in which structural homogeneity is disrupted. Structural defects are found where the honeycomb structure is altered by randomly occurring pentagonal or heptagonal rings (fig. 5b and 5c).[57] It has been shown that structural defects affect the mechanical properties of the graphene.[58] Both the density of defects and their regional arrangements have been shown

---

53 Qiang/Marino/Rui 2009; Marino/Belytschko 2004; Kudin/Scuseria/Yakobson 2001.

54 Ellis/Van der Vies 1991.

55 Ortolani et al. 2012.

56 GPA analysis is a well-established experimental technique aimed at studying strained segments of nanomaterials.

57 Zhang et al. 2010.

58 Lusk/Carr 2008; Grantab/Shenoy/Ruoff 2010; Liu/Yakobson 2010; Rasool et al. 2013.

critical to mechanical properties.[59] For graphene grown on different substrates, such as mica or copper, individually grown crystals meet at a *grain boundary* (GB).

It is known that, in addition to the probable existence of structural defects at any location on grown crystals, the GB region always contains many structural defects. Haider Rasool et al. studied grain boundaries in graphene sheets using HRTEM and found that heptagonal and pentagonal defects are present at the GBs.[60] Rasool et al. observed a variation in carbon bond lengths at the center of the GB regions, where they were either much smaller or much larger compared with the normal length.

A bond length increase or decrease within the GB of the graphene sheet results in a tensile or compressive strain within that region. These strained regions affect the configuration of graphene by creating corrugations or bending in or out of plane of the sheet.

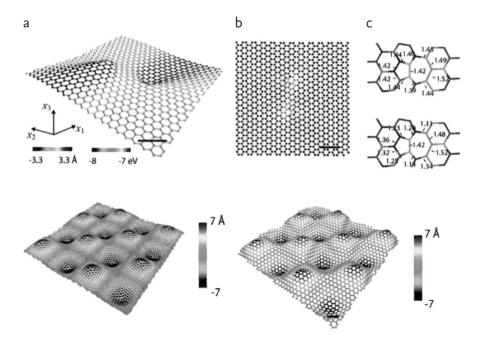

Fig. 5: Configurations of a simulated graphene sheet with dislocation dipole. (a) Perspective view of a region affected by dislocation. (b) Top view of the dislocation affected region. (c) Bond structure around the dislocation core. (d & e) Graphene ruga produced by a periodic placement of disclinations. (d) Continuum modeling and (e) atomistic simulations demonstrating that both methods could simulate the ruga structure of graphene. Scale bar shows 1 nm. Color bar represents variation in height.

59   Wei et al. 2012.
60   Rasool et al. 2013.

In a recent study, Teng Zhang et al. 2014 explored wrinkling and bending in graphene due to *line defects-disclinations* (observed where two or more planes of atoms in a crystal meet).[61] They demonstrated that a significant amount of wrinkling in 2D graphene sheets is likely the result of topological defects caused by disclinations or dislocation of atoms in the graphene lattice, which could be found at the grain boundaries. Furthermore, they examined whether inserting such defects intentionally could help control the wrinkling of graphene sheets to a desired folding or wrinkling configuration.

To answer this question they presented two examples of curved graphene sheets that can be produced and topographically tuned through the variation of topological defects. One of these examples was taken from Alexander Hexemer et al. study.[62] The latter carried out *Monte Carlo* simulations to predict the emergence of disclinations in a lattice of charged particles on a high aspect ratio sinusoidal surface.[63] Hexemer et al. predicted a periodical array of distributed disclination quadruples in a square unit cell creates a wavy graphene (fig. 5d and 5e).

This has been referred to as a *graphene ruga*.[64] Teng Zhang et al.[65] reproduced the graphene ruga through other models and its potential energies were minimized using computer molecular dynamics under periodical boundary conditions. A simple visual comparison between the graphene ruga structure produced by molecular dynamics simulations and the one produced through continuum mechanics calculations demonstrates that purely mechanics-based models (that is continuum models) are able to capture large scale wrinkling induced by structural defects in graphene.

Overall, studies by Zhang et al. have demonstrated that it is feasible to control the structural transition of graphene sheets to 3D structures by manipulation of the position of structural defects. Control over the structural defects introduces new possibilities for the design of application-specific products based on graphene. Changes in configuration introduced through topological defects have been shown to play a critical role in altering the mechanical, thermal and electric properties of graphene sheets.[66]

Apart from the alteration of graphene properties through the introduction of structural defects, Guo Wang et al. have demonstrated that the intentional creation of a line defect

---

61 Zhang et al. 2010.

62 Hexemer et al. 2007.

63 In physics, Monte Carlo methods are usually used for systems with many coupled (interacting) degrees of freedom, such as disordered materials or fluids. Monte Carlo simulations are based on a wide range of different algorithms, which rely on random resampling to obtain numerical results.

64 Zhang/Li/Gao 2014.

65 Zhang et al. 2010

66 Huang et al. 2011.

in graphene offers a way to engineer further wrinkling.[67] Wang et al. carried out molecular simulations of graphene sheets containing one-dimensional periodic defects known as *Stone-Wales defects*.

Anthony Stone and David Wales, studied the icosahedral C_{60} (fullerene) and related structures with a theoretical perspective in 1986.[68] At the time, Stone et al. proposed that C_{60} has many other isomeric structures almost as stable as the icosahedral. The isomers proposed by Stone et al. required certain rearrangements in carbon-carbon bonding, which became known as *Stone-Wales rearrangements*.

A Stone-Wales rearrangement involves a change in π-bonding of two carbon atoms, allowing for a 90 degree rotation with respect to the midpoint of their bond (fig. 6). Inherently, Stone-Wales defects form an out-of-plane bending angle with respect to the longitudinal axis of the sheet, which is known as the defect angle. Wang et al. demonstrated that tuning the defect angle of Stone-Wale defects results in tuning the direction of the graphene wrinkling.

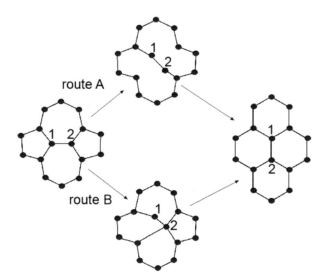

route A

route B

Fig. 6: Stone-Wales rearrangement routes: route A is a concerted mechanism and route B is a stepwise process. According to Stone-Wales rearrangement, two pi-bonded carbon atoms (marked '1' and '2') rotate 90 degree with respect to the midpoint of their bond. The end product of the rearrangement is an energetically more stable structure.

67   Wang et al. 2013.
68   Stone/Wales 1986.

It is known since the early 1990s that the formation of *fullerene*, as a 3D structure,[69] is due to a bottom-up mechanism involving small clusters of carbon atoms.[70] However, it was only in 2010 that Andrey Chuvilin et al. reported the formation of fullerenes *in-situ*.[71] This involved folding of a graphene nanoribbon into a fullerene after exposure to TEM. To understand this process it is important to note that TEM uses an electron beam (e-beam) to produce an image of an ultra-thin sample.

As the electron beam passes through the ultra-thin sample, it interacts with it. When graphene sheets are imaged using TEM, the graphene edges appear to be continuously changing their shapes. Chuvilin et al. observed that the high energy TEM beam of electrons causes fragmentation of larger graphene sheets into smaller ones. As TEM imaging carried on, small graphene flakes tended to transform to fullerene molecules in which the structure remained stable. The authors explained that, for the purpose of transformation of small sheets into fullerene, the creation of defects at the edges of these sheets was important. Defects induced by the electron beam at the edges of the graphene sheets underwent continuous reconstruction, leading to the most stable configuration. Note that the size of the initial 2D sheets is very important for 2D to 3D transformation: for sheets smaller than 60 carbon atoms the strain caused by defects does not suffice to bend the sheet enough to produce a fullerene.

3. Impact of folding on graphene's electronic properties

In this section we review studies concerning the effect of folding on the electronic properties of graphene sheets, especially as it is related to its geometry.

A folded region in a graphene sheet can resemble a fractional nanotube structure, in other words, a hairpin.[72] This implies the electronic properties of a folded graphene sheet are expected to differ from those of a cylindrical carbon nanotube, due to lack of symmetry in the structure of the hairpin-folded graphene.

Ji Feng et al. investigated the effect of a zigzag or armchair structural configuration of carbon bonds on the electronic properties of the folded graphene sheets.[73] The *density functional theory* (DFT) as such does not explain the long range electron correlations

---

69   Icosahedral structure based on carbon rings.
70   Yannoni et al. 1991; Hawkins et al. 1991; Ebbesen/Tabuchi/Tanigaki 1992.
71   Chuvilin et al. 2010.
72   Huang et al. 2009.
73   Feng et al. 2009.

responsible for van der Waals adhesion between bilayers.[74] *Local density approximation* (LDA), on the other hand, employs a high enough level of accuracy to explain the equilibrium interaction potential between graphene layers.[75]

Following DFT-LDA optimizations, Feng et al. demonstrated that the folded graphene sheets with zigzag carbon-carbon bonds at the edge of the fold exhibit a top-bottom separation of 0.52 nm and 0.36 nm at the flat region bilayer separation. In the case of armchair carbon-carbon bonds at the fold edge, the fold top-bottom separation and the flat region bilayer were 0.72 nm and 0.37 nm, respectively. A top-bottom distance variation due to the zigzag and armchair fold edge configurations causes a variation in the scattering of the electrons in the folded region. The zigzag configuration at the fold edge results in a tighter hairpin, which causes more electron scattering and therefore lower conductance (electron mobility) at those folds.

Jhon González et al. investigated the electronic transport properties of folded nanoribbons of graphene sheets by theoretical means.[76] They considered two possibilities for folding: (1) a hairpin-shaped graphene nanoribbon folded on itself with the overlapping flat regions having a 0 degree contact direction (matching lattice), (2) a similarly folded graphene sheet only with overlapping flat regions (top layer) having 60 degree misorientation with the longitudinal axis of the bottom layer.[77]

González et al. demonstrated that, for the hairpin folded graphene nanoribbons having overlapping regions with 0 degree contact direction, conductance is similar to that of a stack of two graphene sheets with matching lattices, as a result of the fact that conductance is affected by the size of the scattering region (within the overlap and arched regions).[78] In the case of a 60 degree contact direction between overlapping regions, a slight mismatch of the sub-lattices disturbed the conductance symmetry within the hairpin-folded graphene nanoribbon. Scattering in the folded regions for both 0 degree and 60 degree caused an overall reduction in conductance of the folded graphene sheets, compared with that of flat single layer sheets. As a consequence, 60 degree folded nanoribbons of graphene have smaller sized scattering regions, giving rise to a conductance value closer to that of unfolded nanoribbons.

---

74 In its simplest definition, density functional theory is a computational quantum-mechanical modeling method used to investigate the electronic structure of atoms and molecules.
75 Girifalco/Hodak 2002.
76 González et al. 2012.
77 Misorientation is the difference in crystallographic orientation between two crystallites.
78 González et al. 2010.

4. Graphene and graphene oxide folding and application-based studies

In this section we describe recent studies on the folding of graphene and graphene oxides with an emphasis on their applications to storage systems. We further present studies on their specific patterning for encoding the folds.

Toby Hallam et al. presented a method termed *Grafold* for printing folded graphene films.[79] Their method is based on using stamps with periodically varying adhesion to produce periodic waves within graphene sheets. Graphene is first transferred onto a patterned elastomeric stamp and then mechanically folded by placing the stamp over the desired substrate and slowly peeling it away. It is important to mention that folds produced in this way in the graphene sheets are more similar to wrinkles.[80] Hallam et al. found that graphene exhibited little strain at these folds (wrinkles) with a charge transport anisotropy within them. They suggested that Grafold could be utilized to produce wrinkles and folds in other 2D sheets such as MoS_2.[81]

Concerning graphene oxide folding, Fei Guo et al. observed that dried folded graphene oxide aggregates undergo a large anisotropic swelling upon rehydration to regain their hydrated unfolded forms.[82] This folding and unfolding was found to be unique to the hydration process, while drying causes a buckling in the structure that stores elastic energy.[83] The authors suggested graphene oxide could constitute a new class of stimuli-responsive materials in nanoelectronic devices or even in controlled drug-release systems.

Chen et al. studied GO-based nanosacks.[84] They demonstrated that, when graphene oxide and a secondary solute are dispersed in water microdroplets within a spray system, a spontaneous segregation into nanosack-cargo nanostructures takes place upon drying. They carried out MD simulations to understand the mechanisms behind this process. They found graphene oxide sheets tend to accumulate at the water/air interface; as drying takes place, the secondary solute is trapped within the graphene oxide sheets in the form of cargo-filled graphene nanosacks. They proposed this process to be scalable – promising for applications where nanomaterials need to be isolated from a specific environment (fig. 7).

---

79   Hallam et al. 2015.
80   See ibid, fig. 2c.
81   Molybdenum disulfide is the inorganic compound with the formula MoS_2.
82   Guo et al. 2011.
83   Elastic energy is the potential mechanical energy stored in the configuration of a material or physical system as work is performed to distort its volume or shape.
84   Chen et al. 2012.

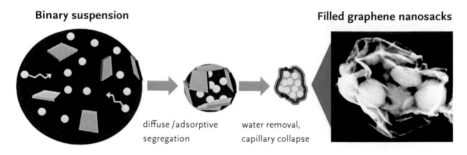

Binary suspension   **Filled graphene nanosacks**

diffuse / adsorptive   water removal,
segregation   capillary collapse

Fig. 7: Colloidal self-assembly of graphene oxide sheets into nanosacks containing the secondary solute. Black and white image shows a scanning electron microscope image of the nanosack filled with nanoparticles.

Soodabeh Movahedi et al. carried out edge modifications of graphene sheets.[85] They proposed a procedure for functionalizing the graphene edge using hyper-branched polyglycerols.[86] It is important to mention the original graphene oxide sheets in their study were synthesized and later reduced to produce reduced graphene oxide sheets. Since hyper-branched polyglycerols are hydrophilic and graphene is hydrophobic, the authors argued that dispersing such a combination in water results in the bending of the hydrophobic graphene portion from a flat surface into a capsule-shaped geometry. The nanocapsules produced using such a method could encapsulate hydrophobic molecules such as doxorubicin, which is a known anti-cancer drug.

Chemical patterning of graphene has served as an important idea in the transformation of 2D flat graphene sheets into more complex 3D configurations. Elias et al. demonstrated that, albeit graphene is one of the most chemically inert materials, it could nevertheless react with atomic hydrogen, thereby transforming it from a highly conductive 2D sheet into an insulating one.[87] They found the hydrogenation process for graphene to be reversible, and conductivity could be restored through merely heating. This is an interesting fact, since it implies that the hydrogenation process affects not only the conductivity but also the 2D flat structure of graphene. This is attributed to disruptions in the honeycomb structure of graphene caused by the hydrogen atoms. Therefore, hydrogenation of graphene sheets at selected areas could be a way of encoding folding (fig. 8). The benefits of this are discussed below.

---

85   Movahedi et al. 2013.
86   Hyper-branched polyglycerols are polymers with a large number of branches, very similar to a tree structure.
87   Elias et al. 2009.

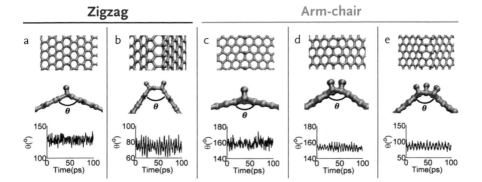

Fig. 8: Top and side view of hydrogen-assisted graphene folding with its resulting fold angles at equilibrium. (a) One-line and (b) two-line hydrogenation are introduced along the zigzag direction of a graphene lattice. (c) One-line, (d) two-line and (e) three-line hydrogenation are introduced along the armchair direction of a graphene lattice. The bottom row plots the variation in folding angle at 300 K over time; variation in the angle of the fold throughout the simulation indicates the stability of the folded substructure.

Zhu et al. carried out systematic MD simulations to demonstrate the feasibility and robustness of *hydrogen-assisted graphene origami* (HAGO).[88] Simulations revealed that, when hydrogenation is present at both sides of a 2D graphene sheet, the structure remains planar; such a fully hydrogenated graphene structure is known as *graphane*.[89] However, when hydrogenation is introduced in a controlled manner along a line formation on any one side of graphene, a net structural distortion can indeed cause folding and bending.

Hydrogen-assisted bending occurs along the hydrogenated line in a certain angle dependent on the number of the hydrogenated rows (fig. 8). It was further demonstrated that an external electric field could further aid in controlling the folding process in HAGO. They attributed the effect of the external electric field to the polarization of carbon atoms in graphene and consequent a change in inter-graphene layer interactions. This demonstrates a decrease of the inter-wall van der Waals energy of adhesion between the graphene layers resulting from an increase in external electric field intensity.

The authors hence propose using graphene-based nanocages as a means of hydrogen storage, while controlling its release through temperature elevation and electric field variation.

---

88   Zhu/Li 2014.
89   Elias et al. 2009; Sofo/Chaudhari/Barber 2007.

Another recent example in using folding for storage purposes is found in the experimental work by Hamid Reza Barzegar et al.[90] they demonstrated a facile solution-based method to insert smaller molecular structures, such as fullerene, C_{60}, into carbon nanotubes. In other words, carbon nanotubes could act as peapods by harboring fullerenes. This was previously reported as a result of simulations by Smith et al., where fullerenes were encapsulated by single-wall carbon nanotubes.[91]

The transformation of carbon nanotubes from a tube into a collapsed and flattened bilayer of graphene, tightly bound with two folded edges, can takes place only when the tube diameter is larger than a certain critical value.[92] Using high resolution TEM, Barzegar et al. observed that the fullerenes are initially packed along the hairpin tube edge and not at the center of the tube. Eventually the nanotube fills up with fullerenes.

Oliver Ochedowski et al. carried out alterations of chemical bonds of graphene sheets at specific locations using *swift heavy ion irradiation* (SHI irradiation).[93] They investigated the effect of high-energy ion beam irradiation on 2D sheets such as graphene and MoS_2. Swift heavy ion radiation specifically interacts and affects the electronic excitation and ionization of a material. Ochedowski et al. proposed two phases in graphene sheets after exposure to SHI radiation.

At the beginning, the graphene sheet bends at the regions exposed to ion radiation and forms a bilayer graphene, a process that can locally increase its mechanical strength. In the second phase, graphene forms a hairpin-folded segment, which affects its chemical reactivity and electron transports at those segments. It was found that, when exposed to the SHI radiation, graphene could transform from 2D into 3D configurations with segments including at least three folds. Such a number of folded segments were not observed, however, in the case of MoS_2 sheets. It is not yet known what is behind the number of folded segments.

Using *scanning force microscopy*, the folded regions in MoS_2 were found to be higher than their graphene counterparts. This was attributed to the inter-layer spacing in MoS_2, which is 0.75 nm and almost twice as high as that of graphene. Folding observed by Ochedowski et al. is believed to be the experimental evidence of defect formation in the exposed areas of graphene and their subsequent self-folding.

---

90   Barzegar et al. 2015.
91   Smith/Monthioux/Luzzi 1998.
92   Chopra et al. 1995; He et al. 2014.
93   Ochedowski et al. 2014.

5. Summary and conclusions

Most simulations and theoretical work done so far emphasized the possibility of folding or self-folding in graphene. There have been studies regarding instances of graphene folding by externally applied forces. However, for a variety of reasons, the guided folding of graphene and graphene oxide has not been demonstrated experimentally yet. Nevertheless, simulation of chemical patterning and modification of graphene sheets was shown to be able to encode the desired folding patterns within the sheets.

Programmed self-assembly of 2D sheets into bulk material with specific properties, or the self-assembly of single sheets into 3D structures, is potentially simple and cheap, hence well suited for advanced applications such as drug delivery or nanostorage. It therefore affords promising possibilities for future industrial production of advanced materials in a way that has advantages over current production methods. Therefore, experimental methods and further investigations in controlling and defining the forces affecting folding at the nanoscale level are needed.

Bibliography

Akhavan, Omid / Ghaderi, Elham / Esfandiar, Ali (2011): *Wrapping Bacteria by Graphene Nanosheets for Isolation from Environment, Reactivation by Sonication, and Inactivation by near-Infrared Irradiation.* In: The Journal of Physical Chemistry B, vol. 115, no. 19, pp. 6279 – 6288.

Arroyo, Marino / Belytschko, Ted (2004): *Finite Crystal Elasticity of Carbon Nanotubes Based on the Exponential Cauchy-Born Rule.* In: Physical Review B, vol. 69, no. 11, pp. 115 – 415.

Barzegar, Hamid Reza / Gracia-Espino, Eduardo / Yan, Aiming / Ojeda-Aristizabal, Claudia / Dunn, Gabriel / Wågberg, Thomas / Zettl, Alex (2015): *C60 / Collapsed Carbon Nanotube Hybrids: A Variant of Peapods.* In: Nano Letters, vol. 15, no. 2, pp. 829 – 834.

Blees, Melina K. / Barnard, Arthur W. / Rose, Peter A. / Roberts, Samantha P. / McGill, Kathryn L. / Huang, Pinshane Y. / Ruyack, Alexander R. / Kevek, Joshua W. / Kobrin, Bryce / Muller, David A. / McEuen, Paul L. (2015): *Graphene Kirigami.* In: Nature, vol. 524, no. 7564, pp. 204 – 207.

Borini, Stefano / White, Richard / Wei, Di / Astley, Michael / Haque, Samiul / Spigone, Elisabetta / Harris, Nadine / Kivioja, Jani / Ryhänen, Tapani (2013): *Ultrafast Graphene Oxide Humidity Sensors.* In: ACS Nano, vol. 7, no. 12, pp. 11166 – 11173.

Brodie, Benjamin C. (1859): *On the Atomic Weight of Graphite.* In: Philosophical Transactions of the Royal Society of London, vol. 149, pp. 249 – 259.

Bunch, Scott J. / Verbridge, Scott S. / Alden, Jonathan S. / van der Zande, Arend M. / Parpia, Jeevak M. / Craighead, Harold G. / McEuen, Paul L. (2008): *Impermeable Atomic Membranes from Graphene Sheets.* In: Nano Letters, vol. 8, no. 8, pp. 2458 – 2462.

Cai, Weiwei / Zhu, Yanwu / Li, Xuesong / Piner, Richard D. / Ruoff, Rodney S. (2009): *Large Area Few-Layer Graphene / Graphite Films as Transparent Thin Conducting Electrodes.* In: Applied Physics Letters, vol. 95, no. 12, 123115.

Cassagneau, Thierry / Fendler, Janos H. (1999): *Preparation and Layer-by-Layer Self-Assembly of Silver Nanoparticles Capped by Graphite Oxide Nanosheets.* In: The Journal of Physical Chemistry B, vol. 103, no. 11, pp. 1789 – 1793.

Castro Neto, Antonio H. / Guinea, F. / Peres, Nuno M. R. / Novoselov, Konstantin S. / Geim, Andre K. (2009): *The Electronic Properties of Graphene.* In: Reviews of Modern Physics, vol. 81, no. 1, pp. 109 – 162.

Chen, Ji / Yao, Bowen / Li, Chun / Shi, Gaoquan (2013): *An Improved Hummers Method for Eco-Friendly Synthesis of Graphene Oxide.* In: Carbon, vol. 64, pp. 225 – 229.

Chen, Sheng / Zhu, Junwu / Wu, Xiaodong / Han, Qiaofeng / Wang, Xin (2010): *Graphene Oxide– Mno2 Nanocomposites for Supercapacitors.* In: ACS nano, vol. 4, no. 5, pp. 2822 – 2830.

Chen, Xiaoming / Zhang, Liuyang / Zhao, Yadong / Wang, Xianqiao / Ke, Changhong (2014): *Graphene Folding on Flat Substrates.* In: Journal of Applied Physics, vol. 116, no. 16, 164301.

Chen, Yantao / Guo, Fei / Jachak, Ashish / Kim, Sang-Pil / Datta, Dibakar / Liu, Jingyu / Kulaots, Indrek / Vaslet, Charles / Jang, Hee Dong / Huang, Jiaxing / Kane, Agnes / Shenoy, Vivek B. / Hurt, Robert Hurt. (2012): *Aerosol Synthesis of Cargo-Filled Graphene Nanosacks.* In: Nano Letters, vol. 12, no. 4, pp. 1996 – 2002.

Cheng-Long, Zhao / Ming, Qin / Wei-Hua, Li / Qing-An, Huang (2011): *Enhanced Performance of a Cmos Interdigital Capacitive Humidity Sensor by Graphene Oxide.* In: Xia, Shanhong / Bao, Minhang / Fan, Long-Sheng (eds.): Proceedings of the 16th International Solid-State Sensors, Actuators and Microsystems Conference, TRANSDUCERS'11. Beijing: IEEE, pp. 1954 – 1957.

Chopra, Nasreen G. / Benedict, Lorin X. / Crespi, Vincent H. / Cohen, Marvin L. / Louie, Steven G. / Zettl, Alex (1995): *Fully Collapsed Carbon Nanotubes.* In: Nature, vol. 377, no. 6545, pp. 135 – 138.

Chung, Chul / Kim, Young-Kwan / Shin, Dolly / Ryoo, Soo-Ryoon / Hong, Byung Hee / Min, Dal-Hee (2013): *Biomedical Applications of Graphene and Graphene Oxide.* In: Accounts of Chemical Research, vol. 46, no. 10, pp. 2211 – 2224.

Chuvilin, Andrey / Kaiser, Ute / Bichoutskaia, Elena / Besley, Nicholas A. / Khlobystov, Andrei N. (2010): *Direct Transformation of Graphene to Fullerene*. In: Nature Chemistry, vol. 2, no. 6, pp. 450–453.

Dill, Ken A. / MacCallum, Justin L. (2012): *The protein-folding problem, 50 years on*. In: Science, vol. 338, no. 6110, pp. 1042–1046.

Dikin, Dmitriy A. / Stankovich, Sasha / Zimney, Eric J. / Piner, Richard D. / Dommett, Geoffrey H. B. / Evmenenko, Guennadi / Nguyen, Son Binh T. / Ruoff, Rodney S. (2007): *Preparation and Characterization of Graphene Oxide Paper*. In: Nature, vol. 448, no. 7152, pp. 457–460.

Dreyer, Daniel R. / Park, Sungjin / Bielawski, Christopher W. / Ruoff, Rodney S. (2010): *The Chemistry of Graphene Oxide*. In: Chemical Society Reviews, vol. 39, no. 1, pp. 228–240.

Ebbesen, Thomas W. / Tabuchi, Junji / Tanigaki, Katsumi (1992): *The Mechanistics of Fullerene Formation*. In: Chemical Physics Letters, vol. 191, no. 3–4, pp. 336–338.

Eilers, Stefan / Rabe, Jürgen P. (2009): *Manipulation of graphene within a scanning force microscope*. In: Physica Status Solidi B, vol. 246, no. 11–12, pp. 2527–2529.

Eilers, Stefan (2013): *Strukturelle und Elektronische Eigenschaften von Nanographen-Graphen-Systemen sowie Schnitt- und Faltverhalten von Graphen*. PhD thesis, Humboldt-Universität zu Berlin.

Elias, Daniel C. / Nair, Rahul Raveendran. / Mohiuddin, Tariq M. G. / Morozov, Sergey V. / Blake, Peter / Halsall, Matthew P. / Ferrari, Andrea C. / Boukhvalov, Danil W. / Katsnelson, Mikhail I. / Geim, Andre K. / Novoselov, Konstantin S. (2009): *Control of Graphene's Properties by Reversible Hydrogenation: Evidence for Graphane*. In: Science, vol. 323, no. 5914, pp. 610–613.

Ellis, R. John / Van der Vies, Saskia M. (1991): *Molecular Chaperones*. In: Annual Review of Biochemistry, vol. 60, no. 1, pp. 321–347.

Fei, L. J. / Debnath, Sujan (2013): *Origami Theory and its Applications: A Literature Review*. In: International Journal of Social, Business, Psychological, Human Science and Engineering, vol. 7, no. 1, pp. 113–117.

Feng, Ji / Li, Wenbin / Qian, Xiaofeng / Qi, Jingshan / Qi, Liang / Li, Ju (2012): *Patterning of Graphene*. In: Nanoscale, vol. 4, no. 16, pp. 4883–4899.

Feng, Ji / Qi, Liang / Huang, Jian Yu / Li, Ju (2009): *Geometric and Electronic Structure of Graphene Bilayer Edges*. In: Physical Review B, vol. 80, no. 16, 165407.

Geim, Andre K. / Novoselov, Konstantin S. (2007): *The Rise of Graphene*. In: Nature Materials, vol. 6, no. 3, pp. 183–191.

Calizo, Irene / Ghosh, S. / Teweldebrhan, D. / Pokatilov, Evghenii P. / Nika, Denis L. / Balandin, Alexander A. / Bao, Wenzhong / Miao, Feng / Lau, C. N. (2008): *Extremely High Thermal Conductivity of Graphene: Prospects for Thermal Management Applications in Nanoelectronic Circuits*. In: Applied Physics Letters, vol. 92, no. 15, 151911.

Girifalco, Louis A. / Hodak, Miroslav (2002): *Van Der Waals Binding Energies in Graphitic Structures*. In: Physical Review B, vol. 65, no. 12, 125404.

González, John W. / Santos, Hernán / Pacheco, Mónica / Chico, Leonor / Brey, Luis (2010): *Electronic Transport through Bilayer Graphene Flakes*. In: Physical Review B, vol. 81, no. 19, 195406.

González, Jhon W. / Pacheco, Mónica / Orellana, Pedro / Brey, Luis / Chico, Leonor (2012): *Electronic Transport of Folded Graphene Nanoribbons*. In: Solid State Communications, vol. 152, no. 15, pp. 1400–1403.

Grantab, Rassin / Shenoy, Vivek B. / Ruoff, Rodney S. (2010): *Anomalous Strength Characteristics of Tilt Grain Boundaries in Graphene*. In: Science, vol. 330, no. 6006, pp. 946–948.

Guo, Fei / Kim, Franklin / Han, Tae Hee / Shenoy, Vivek B. / Huang, Jiaxing / Hurt, Robert H. (2011): *Hydration-Responsive Folding and Unfolding in Graphene Oxide Liquid Crystal Phases*. In: ACS Nano, vol. 5, no. 10, pp. 8019–8025.

Hallam, Toby / Shakouri, Amir / Poliani, Emanuele / Rooney, Aidan P. / Ivanov, Ivan / Potie, Alexis / Taylor, Hayden K. / Bonn, Mischa / Turchinovich, Dmitry / Haigh, Sarah J. / Maultzsch, Janina / Duesberg, Georg S. (2015): *Controlled Folding of Graphene: Grafold Printing*. In: Nano Letters, vol. 15, no. 2, pp. 857–863.

Harrington, Matthew J. / Razghandi, Khashayar / Ditsch, Friedrich / Guiducci, Lorenzo / Rueggeberg, Markus / Dunlop, John W. C. / Fratzl, Peter / Neinhuis, Christoph / Burgert, Ingo (2011): *Origami-like unfolding of hydro-actuated ice plant seed capsules*. In: Nature Communications, vol. 2, no. 337, pp. 1–7.

Hawkins, Joel M. / Meyer, Axel / Loren, Stefan / Nunlist, Rudi (1991): *Statistical Incorporation of $^{13}C_2$ Units into C_{60} (Buckminsterfullerene)*. In: Journal of the American Chemical Society, vol. 113, no. 24, pp. 9394–9395.

He, Maoshuai / Dong, Jichen / Zhang, Kaili / Ding, Feng / Jiang, Hua / Loiseau, Annick / Lehtonen, Juha / Kauppinen, Esko I. (2014): *Precise Determination of the Threshold Diameter for a Single-Walled Carbon Nanotube to Collapse*. In: ACS Nano, vol. 8, no. 9, pp. 9657–9663.

Hexemer, Alexander / Vitelli, Vincenzo / Kramer, Edward J. / Fredrickson, Glenn H. (2007): *Monte Carlo Study of Crystalline Order and Defects on Weakly Curved Surfaces*. In: Physical Review E, vol. 76, no. 5, 051604.

Hofmann, Ulrich / Holst, Rudolf (1939): *Über Die Säurenatur Und Die Methylierung Von Graphitoxyd*. In: Berichte der deutschen chemischen Gesellschaft (A and B Series), vol. 72, no. 4, pp. 754–771.

Huang, Jian Yu / Ding, Feng / Yakobson, Boris I. / Lu, Ping / Qi, Liang / Li, Ju (2009): *In Situ Observation of Graphene Sublimation and Multi-Layer Edge Reconstructions*. In: Proceedings of the National Academy of Sciences, vol. 106, no. 25, pp. 10103–10108.

Huang, Pinshane Y. / Ruiz-Vargas, Carlos S. / van der Zande, Arend M. / Whitney, William S. / Levendorf, Mark P. / Kevek, Joshua W. / Garg, Shivank / Alden, Jonathan S. / Hustedt, Caleb J. / Zhu, Ye / Park, Jiwoong / McEuen, Paul L. / Muller, David A. (2011): *Grains and Grain Boundaries in Single-Layer Graphene Atomic Patchwork Quilts*. In: Nature, vol. 469, no. 7330, pp. 389–392.

Huang, Yonggang / Wu, Jian-Ying / Hwang, Keh-Chih (2006): *Thickness of Graphene and Single-Wall Carbon Nanotubes*. In: Physical Review B, vol. 74, no. 24, 245413.

Hummers, William S. / Offeman, Richard E. (1958): *Preparation of Graphitic Oxide*. In: Journal of the American Chemical Society, vol. 80, no. 6, pp. 1339–1339.

Joshi, Rakesh / Carbone, Paola / Wang, Feng-Chao / Kravets, Vasyl G. / Su, Ying / Grigorieva, Irina V. / Wu, H. A. / Geim, Andre K. / Nair, Rahul Raveendran (2014): *Precise and Ultrafast Molecular Sieving through Graphene Oxide Membranes*. In: Science, vol. 343, no. 6172, pp. 752–754.

Kostarelos, Kostas / Novoselov, Kostya S. (2014): *Exploring the Interface of Graphene and Biology*. In: Science, vol. 344, no. 6181, pp. 261–263.

Kudin, Konstantin N. / Scuseria, Gustavo E. / Yakobson, Boris I. (2001): *C_2f, Bn, and C Nanoshell Elasticity from Ab Initio Computations*. In: Physical Review B, vol. 64, no. 23, 235406.

Lange, Philipp / Dorn, Martin / Severin, Nikolai / Vanden Bout, David A. / Rabe, Jürgen P. (2011): *Single- and Double-Layer Graphenes as Ultrabarriers for Fluorescent Polymer Films*. In: The Journal of Physical Chemistry C, vol. 115, no. 46, pp. 23057–23061.

Le, Nam B. / Woods, Lilia M. (2012): *Folded Graphene Nanoribbons with Single and Double Closed Edges*. In: Physical Review B, vol. 85, no. 3, 035403.

Lee, Changgu / Wei, Xiaoding / Kysar, Jeffrey W. / Hone, James (2008): *Measurement of the Elastic Properties and Intrinsic Strength of Monolayer Graphene*. In: Science, vol. 321, no. 5887, pp. 385–388.

Lee, Sun Hwa / Kim, Hyun Wook / Hwang, Jin Ok / Lee, Won Jun / Kwon, Joon / Bielawski, Christopher W. / Ruoff, Rodney S. / Kim, Sang Ouk (2010): *Three-Dimensional Self-Assembly of Graphene Oxide Platelets into Mechanically Flexible Macroporous Carbon Films*. In: Angewandte Chemie International Edition, vol. 49, no. 52, pp. 10084–10088.

Leenaerts, Ortwin / Partoens, Bart / Peeters, Francois. M. (2008): *Graphene: A Perfect Nanoballoon*. In: Applied Physics Letters, vol. 93, no. 19, 193107.

Li, Hang / Song, Zhuonan / Zhang, Xiaojie / Huang, Yi / Li, Shiguang / Mao, Yating / Ploehn, Harry J / Bao, Yu / Yu, Miao (2013): *Ultrathin, Molecular-Sieving Graphene Oxide Membranes for Selective Hydrogen Separation*. In: Science, vol. 342, no. 6154, pp. 95–98.

Liu, Shaobin / Zeng, Tingying Helen / Hofmann, Mario / Burcombe, Ehdi / Wei, Jun / Jiang, Rongrong / Kong, Jing / Chen, Yuan (2011): *Antibacterial Activity of Graphite, Graphite Oxide, Graphene Oxide, and Reduced Graphene Oxide: Membrane and Oxidative Stress*. In: ACS Nano, vol. 5, no. 9, pp. 6971–6980.

Liu, Yuanyue / Yakobson, Boris I. (2010): *Cones, Pringles, and Grain Boundary Landscapes in Graphene Topology*. In: Nano Letters, vol. 10, no. 6, pp. 2178–2183.

Lusk, Mark T. / Carr, Lincoln D. (2008): *Nanoengineering Defect Structures on Graphene*. In: Physical Review Letters, vol. 100, no. 17, 175503.

Ma, Ming / Grey, François / Shen, Luming / Urbakh, Michael / Wu, Shuai / Liu, Jefferson Zhe / Liu, Yilun / Zheng, Quanshui (2015): *Water Transport inside Carbon Nanotubes Mediated by Phonon-Induced Oscillating Friction*. In: Nature Nanotechnology, vol. 10, no. 8, pp. 692–695.

Meyer, Jannik C. / Geim, Andre K. / Katsnelson, Mikhail I. / Novoselov, Konstantin S. / Booth, Tim J. / Roth, Siegmar (2007): *The Structure of Suspended Graphene Sheets*. In: Nature, vol. 446, no. 7131, pp. 60–63.

Mohanty, Nihar / Fahrenholtz, Monica / Nagaraja, Ashvin / Boyle, Daniel / Berry, Vikas (2011): *Impermeable Graphenic Encasement of Bacteria*. In: Nano Letters, vol. 11, no. 3, pp. 1270–1275.

Movahedi, Soodabeh / Adeli, Mohsen / Fard, Ali Kakanejadi / Maleki, Mahin / Sadeghizadeh, Majid / Bani, Farhad (2013): *Edge-Functionalization of Graphene by Polyglycerol; a Way to Change Its Flat Topology*. In: Polymer, vol. 54, no. 12, pp. 2917–2925.

Nair, Rahul Raveendran / Blake, Peter / Grigorenko, Alexander N. / Novoselov, Konstantin S. / Booth, Tim J. / Stauber, Tobias / Peres, Nuno M. R. / Geim, Andre K. (2008): *Fine Structure Constant Defines Visual Transparency of Graphene*. In: Science, vol. 320, no. 5881, pp. 1308.

Novoselov, Konstantin S. / Geim, Andre K. / Morozov, Sergei V. / Jiang, D. / Zhang, Y. / Dubonos, Sergey V. / Grigorieva, Irina V. / Firsov, Alexandr A. (2004): *Electric Field Effect in Atomically Thin Carbon Films*. In: Science, vol. 306, no. 5696, pp. 666–669.

Ochedowski, Oliver / Bukowska, Hanna / Freire Soler, Victor M. / Brökers, Lara / Ban-d'Etat, Brigitte / Lebius, Henning / Schleberger, Marika (2014): *Folding Two Dimensional Crystals by Swift Heavy Ion Irradiation*. In: Nuclear Instruments and Methods in Physics Research Section B: Beam Interactions with Materials and Atoms, vol. 340, pp. 39–43.

Ortolani, Luca / Cadelano, Emiliano / Veronese, Giulio Paolo / Degli Esposti Boschi, Cristian / Snoeck, Etienne / Colombo, Luciano / Morandi, Vittorio (2012): *Folded Graphene Membranes: Mapping Curvature at the Nanoscale*. In: Nano Letters, vol. 12, no. 10, pp. 5207–5212.

Patra, Niladri / Wang, Boyang / Král, Petr (2009): *Nanodroplet Activated and Guided Folding of Graphene Nanostructures*. In: Nano Letters, vol. 9, no. 11, pp. 3766–3771.

Prada, Elsa / San-Jose, Pablo / Brey, Luis (2010): *Zero Landau Level in Folded Graphene Nanoribbons*. In: Physical Review Letters, vol. 105, no. 10, 106802.

Qiang, Lu / Marino, Arroyo / Rui, Huang (2009): *Elastic Bending Modulus of Monolayer Graphene*. In: Journal of Physics D: Applied Physics, vol. 42, no. 10, 102002.

Rasool, Haider I. / Ophus, Colin / Klug, William S. / Zettl, A. / Gimzewski, James K. (2013): *Measurement of the Intrinsic Strength of Crystalline and Polycrystalline Graphene*. In: Nature Communications, vol. 4, no. 2811, pp. 1–7.

Roy, H. V. / Kallinger, Christian / Marsen, Bjorn / Sattler, Klaus (1998): *Manipulation of Graphitic Sheets Using a Tunneling Microscope*. In: Journal of Applied Physics, vol. 83, no. 9, pp. 4695–4699.

Schmidt, Bernhard V. K. J. / Fechler, Nina / Falkenhagen, Jana / Lutz, Jean-François (2011): *Controlled folding of synthetic polymer chains through the formation of positionable covalent bridges*. In: Nature Chemistry, vol. 3, no. 3, pp. 234–238.

Shu, Lijin / Schlüter, A. Dieter / Ecker, Christof / Severin, Nikolai / Rabe, Jürgen P. (2001): *Extremely Long Dendronized Polymers: Synthesis, Quantification of Structure Perfection, Individualization, and SFM Manipulation*. In: Angewandte Chemie International Edition, vol. 40, no. 24, pp. 4666–4669.

Stephen, Timoshenko S. / Woinowsky, Krieger (1987): *Theory of Plates and Shells*. New York: McGraw-Hill.

Sánchez-Portal, Daniel / Artacho, Emilio / Soler, José M. / Rubio, Angel / Ordejón, Pablo (1999): *Ab Initio Structural, Elastic, and Vibrational Properties of Carbon Nanotubes*. In: Physical Review B, vol. 59, no. 19, pp. 12678–12688.

Smith, Brian W. / Monthioux, Marc / Luzzi, David E. (1998): *Encapsulated C_{60} in Carbon Nanotubes*. In: Nature, vol. 396, no. 6709, pp. 323–324.

Sofo, Jorge O. / Chaudhari, Ajay S. / Barber, Greg D. (2007): *Graphane: A Two-Dimensional Hydrocarbon*. In: Physical Review B, vol. 75, no. 15, 153401.

Sreeprasad, T. Sreenivasan / Berry, Vikas (2013): *How Do the Electrical Properties of Graphene Change with Its Functionalization?*. In: Small, vol. 9, no. 3, pp. 341–350.

Stoller, Meryl D. / Park, Sungjin / Zhu, Yanwu / An, Jinho / Ruoff, Rodney S. (2008): *Graphene-Based Ultracapacitors*. In: Nano letters, vol. 8, no. 10, pp. 3498–3502.

Stone, Anthony J. / Wales, David J. (1986): *Theoretical Studies of Icosahedral C_{60} and Some Related Species*. In: Chemical Physics Letters, vol. 128, no. 5–6, pp. 501–503.

Talapin, Dmitri V. / Shevchenko, Elena V. / Murray, Christopher B. / Titov, Alexey V. / Král, Petr (2007): *Dipole–Dipole Interactions in Nanoparticle Superlattices*. In: Nano Letters, vol. 7, no. 5, pp. 1213–1219.

Tao, Hua-Chao / Fan, Li-Zhen / Mei, Yongfeng / Qu, Xuanhui (2011): *Self-Supporting Si / Reduced Graphene Oxide Nanocomposite Films as Anode for Lithium Ion Batteries*. In: Electrochemistry Communications, vol. 13, no. 12, pp. 1332–1335.

Tarsicio, Beléndez / Cristian, Neipp / Augusto, Beléndez (2002): *Large and Small Deflections of a Cantilever Beam*. European Journal of Physics, vol. 23, no. 3, 371.

Tersoff, Jerry (1988): *New Empirical Approach for the Structure and Energy of Covalent Systems*. In: Physical Review B, vol. 37, no. 12, pp. 6991–7000.

Tu, Zhan-chun / Ou-Yang, Zhong-can (2002): *Single-Walled and Multiwalled Carbon Nanotubes Viewed as Elastic Tubes with the Effective Young's Moduli Dependent on Layer Number*. In: Physical Review B, vol. 65, no. 23, 233407.

Wang, C. Guo / Lan, Lan / Liu, Yuanpeng / Tan, Huifeng (2013): *Defect-Guided Wrinkling in Graphene*. In: Computational Materials Science, vol. 77, pp. 250–253.

Wang, Zhong-Li / Xu, Dan / Xu, Ji-Jing / Zhang, Lei-Lei / Zhang, Xin-Bo (2012): *Graphene Oxide Gel-Derived, Free-Standing, Hierarchically Porous Carbon for High-Capacity and High-Rate Rechargeable $Li-O_2$ Batteries*. In: Advanced Functional Materials, vol. 22, no. 17, pp. 3699–3705.

Wei, Yujie / Wu, Jiangtao / Yin, Hanqing / Shi, Xinghua / Yang, Ronggui / Dresselhaus, Mildred (2012): *The Nature of Strength Enhancement and Weakening by Pentagon–Heptagon Defects in Graphene*. In: Nature Materials, vol. 11, no. 9, pp. 759–763.

Wissler, Mathis (2006): *Graphite and Carbon Powders for Electrochemical Applications*. In: Journal of Power Sources, vol. 156, no. 2, pp. 142–150.

Wu, L. / Qu, P. / Zhou, R. / Wang, B. / Liao, S. (2015): *Green Synthesis of Reduced Graphene Oxide and Its Reinforcing Effect on Natural Rubber Composites*. In: High Performance Polymers, vol. 27, no. 4, pp. 486–496.

Xianhong, Meng / Ming, Li / Zhan, Kang / Xiaopeng, Zhang / Jianliang, Xiao (2013): *Mechanics of Self-Folding of Single-Layer Graphene*. In: Journal of Physics D: Applied Physics, vol. 46, no. 5, 055308.

Yang, Shubin / Feng, Xinliang / Ivanovici, Sorin / Müllen, Klaus (2010): *Fabrication of Graphene-Encapsulated Oxide Nanoparticles: Towards High-Performance Anode Materials for Lithium Storage*. In: Angewandte Chemie International Edition, vol. 49, no. 45, pp. 8408–8411.

Yannoni, Costantino S. / Bernier, Patrick P. / Bethune, Donald S. / Meijer, Gerard / Salem, Jesse R. (1991): *NMR Determination of the Bond Lengths in C_{60}*. In: Journal of the American Chemical Society, vol. 113, no. 8, pp. 3190–3192.

Zhang, Jiong / Xiao, Jianliang / Meng, Xianhong / Monroe, Carolyn / Huang, Yonggang / Zuo, Jian-Min (2010): *Free Folding of Suspended Graphene Sheets by Random Mechanical Stimulation*. In: Physical Review Letters, vol. 104, no. 16, 166805.

Zhang, Liuyang / Zeng, Xiaowei / Wang, Xianqiao (2013): *Programmable Hydrogenation of Graphene for Novel Nanocages*. In: Scientific Reports, vol. 3, 3162.

Zhang, Teng / Li, Xiaoyan / Gao, Huajian (2014): *Defects Controlled Wrinkling and Topological Design in Graphene*. In: Journal of the Mechanics and Physics of Solids, vol. 67, pp. 2–13.

Zhang, Teng / Li, Xiaoyan / Gao, Huajian (2014): *Designing Graphene Structures with Controlled Distributions of Topological Defects: A Case Study of Toughness Enhancement in Graphene Ruga*. In: Extreme Mechanics Letters, vol. 1, pp. 3–8.

Zhou, Weiwei / Zhu, Jixin / Cheng, Chuanwei / Liu, Jinping / Yang, Huanping / Cong, Chunxiao / Guan, Cao / Jia, Xingtao / Fan, Hong Jin / Yan, Qingyu / Li, Chang Ming / Yu, Ting (2011): *A General Strategy toward Graphene@Metal Oxide Core-Shell Nanostructures for High-Performance Lithium Storage*. In: Energy and Environmental Science, vol. 4, no. 12, pp. 4954–4961.

Zhuang, Wei / Kasëmi, Edis / Ding, Yi / Kröger, Martin / Schlüter, A. Dieter / Rabe, Jürgen P. (2008): *Self-Folding of Charged Single Dendronized Polymers*. In: Advanced Materials, vol. 20, no. 17, pp. 3204–3210.

Zhu, Mingshan / Chen, Penglei / Liu, Minghua (2011): *Graphene Oxide Enwrapped Ag/AgX (X = Br, Cl) Nanocomposite as a Highly Efficient Visible-Light Plasmonic Photocatalyst*. In: ACS Nano, vol. 5, no. 6, pp. 4529–4536.

Zhu, Shuze / Li, Teng (2014): *Hydrogenation-Assisted Graphene Origami and Its Application in Programmable Molecular Mass Uptake, Storage, and Release*. In: ACS Nano, vol. 8, no. 3, pp. 2864–2872.

Mohammad Fardin Gholami
Email: *gholami@physik.hu-berlin.de*

Image Knowledge Gestaltung. An Interdisciplinary Laboratory.
Cluster of Excellence Humboldt-Universität zu Berlin.
Sophienstrasse 22a, 10178 Berlin, Germany.

Department of Physics and IRIS Adlershof, Humboldt-Universität zu Berlin.
Newtonstr. 15, 12489 Berlin, Germany.

Nikolai Severin, Jürgen P. Rabe
Department of Physics and IRIS Adlershof, Humboldt-Universität zu Berlin.
Newtonstr. 15, 12489 Berlin, Germany.

Credits of Images

Krauthausen: *The Dimensionality of Writing in French Structuralism (1966–1972)*

Fig. 1: Laufer 1988, 69. © Bernhard Laufer.

Fig. 2: Diderot / le Rond d'Alembert 1751–1780, Planche VI. Source: *www.planches.eu* [last access: 8 January 2016].

Fig. 3: Photo: Friedrich Forssman. © Friedrich Forssman.

Seppi: *Simply complicated: Thinking in folds*

Fig. 1: Deleuze 1988, 7. © Les Editions de Minuit, Paris 1988.

Fig. 2: Klee 1953, 6. © Praeger New York.

Fig. 3: © bpk / Scala / Florenz, Santa Maria Novella 2016.

Fig. 4: © Museo del Prado, Madrid 2016.

Fig. 5: © bpk / RMN – Grand Palais / Jean Schormans / Paris, Musée du Louvre 2016.

Fig. 6, 7: © VG Bild-Kunst, Bonn 2016.

Fig. 8: © bpk / Scala / Rom, Galleria Nazionale d'Arte Antica, Palazzo Corsini 2016.

Fig. 9: © bpk / Alinari Archives / Bencini Raffaello / S. Tomé, Toledo 2016.

Fig. 10: © Cameraphoto / Scala, Florenz 2016.

Fig. 11: Lærke 2015, 16. © BJHP / Laerke 2015.

Blümle: *Infinite Folds: El Greco and Deleuze's Operative Function of the Fold*

Fig. 1, 3: © Museo Thyssen-Bornemisza / Scala, Florence 2016.

Fig. 2, 5–7: © Museo Thyssen-Bornemisza / Scala, Florence 2016. Graphic editing: Claudia Blümle, Berlin | *Bild Wissen Gestaltung* 2016.

Fig. 4: © Museo Nacional del Prado. Photo: MNP / Scala, Florence 2016.

Ferrand/Peysson: *Versal Unfolding: How a Specific Folding Can Turn Crease and Tear into Transversal Notions*

Fig. 1 – 8: © Photo: Emmanuel Ferrand and Dominique Peysson, Paris | Institut de Mathématiques de Jussieu-Paris Rive Gauche, Université Pierre and Marie Curie and École Nationale Supérieure des Arts décoratifs 2015.

Schramke: *3D Code: Folding in the Architecture of Peter Eisenman*

Fig. 1: © Lucio Nardi 2015.

Fig. 2: Ungers 1985.

Fig. 3: © Graphics: Diller Scofidio + Renfro: Computer model for the Blur pavilion, Suisse, 2002.

Fig. 4: Kammer/Hinrichs 1993, 46.

Friedman/Krausse: *Folding and Geometry: Buckminster Fuller's Provocation of Thinking*

Fig. 1, 2, 6: © Estate of R. Buckminster Fuller 1997.

Fig. 3: Kepler 1611, 14.

Fig. 4: (a) Fuller 1975, Figure 413.01(A); **(b)** Graphic: Michael Friedman, Berlin | *Bild Wissen Gestaltung* 2015.

Fig. 5: © Graphic: Michael Friedman, Berlin | *Bild Wissen Gestaltung* 2015.

Fig. 7: Screenshots: Fuller 1975b.

Guiducci/Dunlop/Fratzl: *An Introduction into the Physics of Self-folding Thin Structures*

Fig. 1, 5: © Photo: Lorenzo Guiducci, Berlin | *Bild Wissen Gestaltung* 2015.

Fig. 2: Adapted from the original, figure released in the public domain.

Fig. 3: (a), (b) above © Photo and Graphic: Lorenzo Guiducci, Berlin | *Bild Wissen Gestaltung* 2015; **(b) below** © Photo: Elbaum et al. 2007, fig. 4, adapted from the original, used with permission from The American Association for the Advancement of Science.

Fig. 4: (a – c) © Dunlop 2016, fig. 3 – 2; **(d)** Photo: John W. C. Dunlop, Potsdam | MPI für Kolloid- und Grenzflächenforschung 2014; **(e)** Photo: Khashayar Razghandi, Potsdam | MPI für Kolloid- und Grenzflächenforschung 2013; **(f)** © Kim et al. 2012, fig. 1, adapted from the original, used with permission from The American Association for the Advancement of Science.